SpringerBriefs in Computer Science

Series Editors
Stan Zdonik
Peng Ning
Shashi Shekhar
Jonathan Katz
Xindong Wu
Lakhmi C. Jain
David Padua
Xuemin (Sherman) Shen
Borko Furht
V.S. Subrahmanian
Martial Hebert
Katsushi Ikeuchi
Bruno Siciliano

T0211842

For further volumes:
http://www.springer.com/series/10028

Saman Atapattu • Chintha Tellambura • Hai Jiang

Energy Detection for Spectrum Sensing in Cognitive Radio

Springer

Saman Atapattu
Department of Electrical
  and Computer Engineering
University of Alberta
Edmonton, AB, Canada

Chintha Tellambura
Department of Electrical
  and Computer Engineering
University of Alberta
Edmonton, AB, Canada

Hai Jiang
Department of Electrical
  and Computer Engineering
University of Alberta
Edmonton, AB, Canada

ISSN 2191-5768   ISSN 2191-5776 (electronic)
ISBN 978-1-4939-0493-8   ISBN 978-1-4939-0494-5 (eBook)
DOI 10.1007/978-1-4939-0494-5
Springer New York Heidelberg Dordrecht London

Library of Congress Control Number: 2014930296

Printed on acid-free paper

Springer is part of Springer Science+Business Media (www.springer.com)

Preface

Spectrum sensing is critically important for cognitive radio, an emerging solution to the spectrum congestion and low usage of licensed spectrum. Energy detection is a promising low-complexity and low-cost spectrum sensing technique. Its performance analysis has been revisited extensively in the recent literature. This book thus aims at a comprehensive summary of recent research on energy detection for spectrum sensing in cognitive radio networks. This book is for researchers and engineers in both industry and academia who would like to know more about applications of energy detection.

After introducing cognitive radio and spectrum sensing techniques in Chap. 1, we discuss the basics of conventional energy detection in detail in Chap. 2. To improve conventional energy detection, many alternative energy detection techniques have been developed, which are described in Chap. 3. The common performance measures of energy detector are described in Chap. 4. Finally, Chap. 5 deals with diversity and cooperative spectrum sensing techniques which can significantly improve energy detection performance.

We would like to thank Dr. Xuemin (Sherman) Shen, for his help in publishing this monograph.

Edmonton, AB, Canada   Saman Atapattu
Edmonton, AB, Canada   Chintha Tellambura
Edmonton, AB, Canada   Hai Jiang
December, 2013

Contents

Acronyms

ADC	Analog-to-digital converter
AUC	Area under the receiver operating characteristic curve
AWGN	Additive white Gaussian noise
BCED	Blindly combined energy detection
BER	Bit error rate
BSC	Binary symmetric channel
CAUC	Complementary AUC
CCI	Co-channel interference
CDF	Cumulative distribution function
CLT	Central limit theorem
CSCG	Circularly symmetric complex Gaussian
CSI	Channel-state information
DSA	Dynamic spectrum access
EGC	Equal gain combining
ENP	Estimated noise power
FCC	Federal Communications Commission
FDMA	Frequency division multiple access
GGN	Generalized Gaussian noise
GMN	Gaussian mixture noise
ICA	Independent component analysis
i.i.d.	Independent and identically distributed
LLR	Log-likelihood ratio
LTE	Long term evolution
MG	Mixture gamma
MGF	Moment generating function
MIMO	Multiple-input multiple-output
MRC	Maximal ratio combining
OFDM	Orthogonal frequency division multiplexing
PDF	Probability density function
PSK	Phase shift keying
ROC	Receiver operating characteristic

SC	Selection combining
SLC	Square-law combining
SLS	Square-law selection
SNR	Signal-to-noise ratio
SPRT	Sequential probability ratio test
SUN	Smart utility networks
TDMA	Time division multiple access
TV	Television
WiMAX	Worldwide interoperability for microwave access
WLAN	Wireless local area network
WRAN	Wireless regional area network

Chapter 1
Introduction

1.1 Wireless Communications

In recent decades, the market for wireless devices and networks has boosted an unprecedented growth. This growth has led to numerous wireless services and applications. Consequently, regulatory agencies in different countries thus allocate (licensed) chunks of spectrum to different wireless services. For instance, the radio spectrum allocated for different applications is shown in Fig. 1.1.

These emerging and relentless growth of wireless networks have increased the demand for spectrum. To meet the rising demand, effective utilization of spectrum is the goal of the following technologies:

- Multiple-input multiple-output (MIMO) communications: MIMO systems allow higher data throughput without additional spectrum usage by spreading the same total transmit power over the antennas, which improves spectral efficiency. For example, IEEE 802.11n (Wi-Fi) uses MIMO to achieve the maximum data rate up to 600 Mbps at 2.4 GHz [27]. Different MIMO systems include single-user and multi-user MIMO. For a single-user MIMO network with $n_T (\geq 1)$ transmit and $n_R(\geq 1)$ receive antennas, the capacity of a single link increases linearly with $\min(n_T, n_R)$. This increase also motivates a *multi-user MIMO* network which achieves the similar capacity scaling when an access point with n_T transmit antennas communicates with n_R users [31]. Larger diversity gains can be achieved when each user has multiple antennas. Multi-user MIMO will be implemented in IEEE 802.11ac (in early 2014) which enables multi-station wireless local area network (WLAN) with throughput of at least 1 Gbps [20]. In addition, *massive MIMO* using large-scale antenna arrays is capable of shrinking the cell size and reducing the transmit power and overhead for channel training (if channel reciprocity is exploited) [29].

S. Atapattu et al., *Energy Detection for Spectrum Sensing in Cognitive Radio*,
SpringerBriefs in Computer Science, DOI 10.1007/978-1-4939-0494-5_1,
© The Author(s) 2014

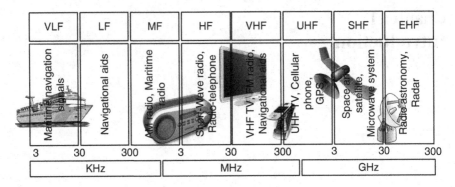

Fig. 1.1 Radio spectrum allocated for wireless communications

- Cooperative communications: Cooperative radio transmissions improve both reliability and data rate by exploiting distributed spatial diversity in a multi-user environment. Cooperative techniques include relaying, cooperative MIMO, and multi-cell MIMO. Single-user or multi-user relaying facilitates the signal transmission between the source and the destination utilizing less power [1,25]. Cooperative MIMO, which forms a distributed antenna system employing antennas of different users, is effective for poor line-of-sight propagation and for cell-edge users. Cooperative MIMO utilizes the advantages of both MIMO and cooperative communications techniques [28]. Further, the larger number of users/antennas in MIMO networks and the universal frequency reuse (e.g., in long term evolution (LTE)-advanced) cause high levels of co-channel interference. Such interference can be mitigated by multiple cells cooperating, referred to as *multi-cell MIMO* [12]. The cooperation among multi-cell base stations can be established via high-capacity wired backhaul links.
- Heterogeneous networks: These utilize a diverse set of base stations in different cells to improve spectral efficiency per unit area, which is necessary to support increasing node density and cell traffic in mobile networks. However, since traditional cellular networks are almost at their peak capacities, to meet future growth, heterogeneous designs have been envisaged. These include a disparate mix of base stations and cells such as lower-power base station in pico-cells (250 mW–2 W) and femto cells (100 mW or less), and high-speed WLANs. A user may be switched among the macro-cells, pico-cells, femto cells, and the WLANs [6].

Despite these advanced technologies, spectrum scarcity continues to create challenges for the Federal Communications Commission (FCC) in the United States and regulatory agencies in other countries. A promising solution is *cognitive radio* technology [13,23].

1.2 Cognitive Radio

What has primarily motivated cognitive radio technology, an emerging novel concept in wireless access, is apparent spectrum under-utilization. The experiments by FCC show that at any given time and location, much (between 80% and 90%) of the licensed spectrum is underutilized [13]. Such temporarily unused spectrum slots are called spectrum holes, resulting in spectral inefficiency. Thus not only is spectrum usage low in some licensed bands, but also true scarcity of radio spectrum compounds the problem. Consequently, the growth of wireless applications may be hindered.

Key features of a cognitive radio transceiver thus include radio environment awareness and spectrum intelligence. The latter refers to an ability to learn the spectrum environment and adapt transmission parameters [13]. For instance, two types of cognitive radio networks are distinguished based on the spectrum bands [2]:

- On unlicensed bands: These include ISM (industrial, scientific and medical) bands such as 902–928 MHz, 2.4–2.5 GHz, and 5.725–5.875 GHz. ISM bands are also shared with non-ISM applications, e.g., Bluetooth: 2.402–2.48 GHz; IEEE 802.11/WiFi: 2.45 and 5.8 GHz bands; and IEEE 802.15.4, ZigBee and other personal area networks: 915 MHz and 2.45 GHz bands. These bands can be utilized by cognitive radio.
- On licensed bands: The spectrum is licensed into different applications, e.g., aeronautical and maritime communications: 300–535 kHz; AM radio: 535 kHz and 1.605 MHz; and LTE-North America: 700 MHz, 800 MHz, 1.9 GHz, 1.7/2.1 GHz and 2.6 GHz. Since significant under-utilization of licensed spectrum can occur, cognitive radio aims to overcome this. For instance, the wireless regional area network (WRAN) standard operates in unused television (TV) channels in 698–806 MHz.

1.2.1 Applications of Cognitive Radio

Some applications of cognitive radio technology are as follows [26, 34].

- Smart grid networks: Currently, there is a need to transform the traditional power grids to smart grids with smart meters for billing. Since these meters require information exchange between a premises and a network gateway (with a distance from a few hundred meters to a few kilometers), a reliable communication system is required. Conventional options such as power line communications support only low data rate and shorter distance, and the cellular networks may not have enough bandwidth. Therefore, IEEE 802.15.4g Smart Utility Networks (SUN) Task Group, which provides a global standard for smart grid networks, seeks cognitive radio solutions that offer advantages in terms of bandwidth, coverage range and overhead [32].

- Public safety networks: Public services such as police, fire, and medical services largely use fixed wireless spectrum, which can be highly congested in emergencies, resulting in a delayed response to victims. In addition to fixed allocated radio spectrum, public services can also use unlicensed spectrum or the TV white spaces (i.e., frequency slots unused by TV broadcasters) to ensure sufficient capacity to achieve effective timely communications [34]. For example, since the US Department of Homeland Security considers the National Emergency Communications Plan, cognitive radio techniques may help to improve the quality of service of such public services [8].
- Cellular networks: Since current cellular networks are operating at peak traffic levels, TV white spaces may be available for the cellular operators in future to use cognitive radio techniques [5]. However, integration of cognitive radio technologies and cellular networks, e.g., LTE and Worldwide Interoperability for Microwave Access (WiMAX), remains to be investigated.

1.2.2 Dynamic Spectrum Access

Dynamic spectrum access (DSA) is defined as a technique by which the operating spectrum of a radio network can be selected dynamically from the available spectrum [16]. DSA can be applied in cognitive radio networks, which has a hierarchical access structure with licensed users (*primary users*) and unlicensed users (*secondary users* or *cognitive users*) (Fig. 1.2).

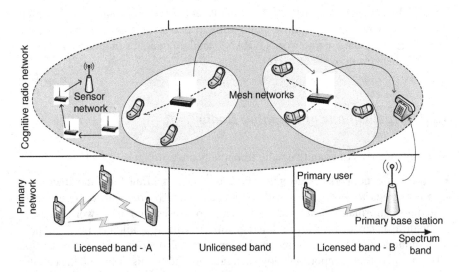

Fig. 1.2 Cognitive radio network architecture

The basic idea of DSA is to open licensed spectrum to secondary users while limiting the interference perceived by primary users [35]. DSA requires opportunistic spectrum sharing, which is implemented via two strategies [41].

1. *Spectrum overlay* does not necessarily impose severe constraints on the transmit power of secondary users, but rather on their transmission time. Consequently, a secondary user accesses a spectrum hole assigned via DSA.
2. *Spectrum underlay* imposes strict constraints on the transmit power of secondary users. Their transmit power is thus low enough to be regarded as noise by primary users. Both primary and secondary users may thus transmit simultaneously in the same spectrum band.

While secondary overlay users opportunistically access spectrum holes, if and when primary users become active by transmitting and receiving, secondary users must detect these transmissions reliably, immediately vacate the channel, and find other spectrum holes for continuing communication. One of the most important tasks is thus to identify the spectrum holes. However, this task is not required in the underlay cognitive radio.

1.3 Spectrum Sensing

Spectrum holes, i.e., available channels, must be sensed for opportunistic spectrum access. Their successful detection allows overlay user access [36]. Since spectrum holes are due to idle state (i.e., no signal transmission) of primary users, to identify spectrum holes, the secondary users must detect the absence of primary signals in a given frequency slot. This task can be viewed as a binary hypothesis testing problem in which Hypothesis 0 ($\mathcal{H}_0$) and Hypothesis 1 ($\mathcal{H}_1$) are the primary signal absence and the primary signal presence, respectively [19]. Spectrum holes are thus identified when $\mathcal{H}_0$ is true.

While signal detection has been common in traditional networks, the following challenges arise with its use for cognitive radio spectrum sensing.

- A much more reliable detector (than those in traditional networks) is required, since any missed-detection results in secondary transmissions which will interfere with the undetected active primary users. Thus, IEEE 802.22 standard specifies 90% accurate detection capability.
- A much wider spectrum bandwidth needs to be sensed to identify as many as possible spectrum holes, e.g., one TV channel bandwidth (6–7 MHz) is not sufficient for 4G mobile communications applications using up to 20 MHz bandwidth. Further, since different frequency bands experience different signal propagation characteristics, design of detection algorithms and their performance analysis are challenging.
- Over a large variety of transmission environments, signal transmission from multiple licensed wireless nodes must be detected. Different applications may have different populations of users, with different mobility patterns, which have a great impact on the signal detection.

Further, although signal detection has been well investigated for traditional wireless networks, the same techniques may fail in cognitive radio networks, because the available knowledge of the primary network parameters at secondary users is limited, and computational complexity and implementation cost are also main factors.

1.4 Spectrum Sensing Techniques

Four major spectrum sensing techniques are energy detection, matched filter, cyclostationary feature detection, and eigenvalue detection.

- Energy detection: This measures the received signal energy within the pre-defined bandwidth and time period. The measured energy is then compared with a threshold to determine the status (presence/absence) of the transmitted signal. Not requiring channel gains and other parameter estimates, the energy detector has low implementation cost. However, its performance degrades with high noise uncertainty and high background interference [22].
- Matched filter: This detector performs coherent operations, and thus requires perfect knowledge of the transmitted signal and the channel response. It is the optimal detector (in the Neyman-Pearson sense) that maximizes the signal to noise ratio (SNR) in the presence of additive noise. Since it requires perfect timing and synchronization at both physical and medium access control layers, computational complexity is high. Its performance decreases dramatically when channel response changes rapidly. In addition, when it is employed to sense spectrum holes, the presence of multiple primary user signals over the same bandwidth simultaneously can impact the accuracy of its decisions. This problem may be mitigated by having a dedicated matched filter structure for each primary signal [2, 4, 30]. However, the resulting complexity issues may preclude the use of matched filter for dynamic and opportunistic spectrum sensing.
- Cyclostationary feature detection: If periodicity properties are introduced intentionally to primary user signals, the statistical parameters of received signal such as mean and autocorrelation may vary periodically. Such statistical periodicity is exploited in cyclostationary detection. One possible way of extracting cyclostationary properties is by using the input-output spectral correlation density or cyclic spectrum. Since noise signal does not have any cyclostationary or periodicity property, this method allows the determination of signal presence/absence readily. While this detector is able to distinguish among the primary user signals, secondary user signals, or interference, it needs a higher sampling rate to get a sufficient number of samples, which increases the computational complexity. In addition, when there are frequency offset and sample timing error, the spectral correlation density may be weak, thus largely affecting detection performance [10, 11, 21, 37].

- Eigenvalue detection: The ratio of the maximum (or the average) eigenvalue to the minimum eigenvalue of the covariance matrix of the received signal vector is compared with a threshold to detect the absence or the presence of the primary signal. However, if the correlation of the primary signal samples is zero (e.g., primary signal appears as white noise), eigenvalue detection may fail, a very rare event. This detector has the advantage of not requiring the knowledge of the primary signal parameters or the propagation channel conditions. The main drawback is computational complexity of covariance matrix computation and the eigenvalue decomposition. The threshold selection is challenging as well [17, 38]. Using the similar concept, a sample autocorrelation-based spectrum sensing technique is also introduced in [24].

Apart from these main spectrum sensing techniques, there are some alternatives:

- The Anderson-Darling sensing is a non-parametric hypothesis testing problem (a goodness-of-fit testing problem) [33]. This technique tests whether or not the observed samples are drawn independently from the noise distribution. If the test does not satisfy the properties of the noise distribution, the detector decides on signal presence. Thus, any assumption/knowledge of the transmitted signal is unnecessary.
- The Kolmogorov-Smirnov test [39] is also a non-parametric method to measure the goodness of fit. This technique computes the empirical cumulative distribution function, and compares it with the known empirical cumulative distribution function of the noise samples.
- By introducing the received signal into a dynamic stochastic resonance system, the SNR of primary signals can be increased. This technique is introduced for the spectrum sensing in [14, 15] in which the spectral power of the primary user signal is amplified, and thus, the received SNR is improved. The stochastic resonance spectrum sensing is useful at low SNR.
- For Gaussian distributed signals, the cumulants (which can be expressed in term of moments) higher than second order are zero [9]. This property can be used to differentiate non-Gaussian signals from Gaussian signals. Since the presence of the primary user induces non-Gaussian property in the received signal, a single and multiple cumulants based spectrum sensing technique is proposed in [18]. However, the detector may fail if the received signals in both hypotheses are Gaussian.
- Diffusion detection schemes, which use adaptive detection algorithms, can track changes of the parameters associated with transmitters, receivers and channels. Recursive least square and least-mean square algorithms are two diffusion spectrum sensing schemes [3]. Diffusion techniques can improve the robustness specially when the network parameters vary rapidly.
- Most detectors also estimate noise power. If this estimate has errors, the performance is poor. Thus, as a solution to noise uncertainty, a frequency-domain entropy-based spectrum sensing technique is proposed in [40]. As another solution to noise uncertainty, blind spectrum sensing algorithms do not require

the knowledge of the signal, the noise power or the channel, and thus, it may have low complexity. Such an algorithm in which the received signal is over-sampled is introduced in [7].

Among the aforementioned spectrum sensing techniques, energy detection has the advantages of low complexity and low cost. Thus, it is the focus of this monograph. The conventional energy detector and system model are discussed in Chap. 2. Chapter 3 briefly explains alternative energy detection techniques. Different performance metrics of the energy detection are discussed in Chap. 4. The energy detection based diversity techniques and cooperative spectrum sensing networks are discussed in Chap. 5.

References

1. Atapattu, S., Jing, Y., Jiang, H., Tellambura, C. (2013) Relay selection and performance analysis in multiple-user networks. J on Selected Areas in Communications **31**(8): 1517–1529.
2. Cabric, D., Mishra, S. M., Brodersen, R. W. (2004) Implementation issues in spectrum sensing for cognitive radios. In: Asilomar Conference on Signals, Systems and Computers, Pacific Grove, 7–10 Nov 2004.
3. Cattivelli, F. S., Sayed, A. H. (2011) Distributed detection over adaptive networks using diffusion adaptation. IEEE T Signal Processing **59**(5): 1917–1932.
4. Chen, H. S., Gao, W., Daut, D. G. (2007) Signature based spectrum sensing algorithms for IEEE 802.22 WRAN. In: Proceedings of IEEE International Conference on Communications (ICC), Glasgow, 24–28 June 2007.
5. Connecting-America (2010) The National Broadband Plan. http://download.broadband.gov/plan/national-broadband-plan.pdf.
6. Damnjanovic, A., Montojo, J., Wei, Y., Ji, T., Luo, T., Vajapeyam, M., Yoo, T., Song, O., Malladi, D. (2011) A survey on 3GPP heterogeneous networks. IEEE Wireless Communications **18**(3): 10–21.
7. De, P., Liang, Y. C. (2008) Blind spectrum sensing algorithms for cognitive radio networks. IEEE T on Vehicular Technology **57**(5): 2834–2842.
8. Doumi, T. L. (2006) Spectrum considerations for public safety in the United States. IEEE Communications M **44**(1): 30–37.
9. Fan, H., Meng, Q., Zhang, Y., Feng, W. (2006) Feature detection based on filter banks and higher order cumulants. In: Proceedings of IEEE International Conference on Information and Acquisition (ICIA), Colombo, 15–17 Dec 2006.
10. Gardner, W. A. (1988) Signal interception: a unifying theoretical framework for feature detection. IEEE T on Communications **36**(8): 897–906.
11. Gardner, W. A. (1991) Exploitation of spectral redundancy in cyclostationary signals. IEEE Signal Processing M **8**(2): 14–36.
12. Gesbert, D., Hanly, S., Huang, H., Shitz, S. S., Simeone, O., Yu, W. (2010) Multi-cell MIMO cooperative networks: A new look at interference. IEEE J on Selected Areas in Communications **28**(9): 1380–1408.
13. Haykin, S. (2005) Cognitive radio: Brain-empowered wireless communications. IEEE J on Selected Areas in Communications **23**(2): 201–220.
14. He, D. (2013) Chaotic stochastic resonance energy detection fusion used in cooperative spectrum sensing. IEEE T on Vehicular Technology **62**(2): 620–627.
15. He, D., Lin, Y., He, C., Jiang, L. (2010) A novel spectrum-sensing technique in cognitive radio based on stochastic resonance. IEEE T on Vehicular Technology **59**(4): 1680–1688.

16. IEEE1900.1-2008 IEEE standard definitions and concepts for dynamic spectrum access: Terminology relating to emerging wireless networks, system functionality, and spectrum management. http://standards.ieee.org/findstds/standard/1900.1-2008.html.

17. Juang, B. H., Li, G. Y., Ma, J. (2009) Signal processing in cognitive radio. Proceedings of the IEEE 97(5): 805–823.

18. Jun, W., Guangguo, B. (2010) Spectrum sensing in cognitive radios based on multiple cumulants. IEEE Signal Processing Letters 17(8): 723–726.

19. Liang, Y. C., Zeng, Y., Peh, E. C. Y., Hoang, A. T. (2008) Sensing-throughput tradeoff for cognitive radio networks. IEEE T on Wireless Communications 7(4): 1326–1337.

20. Liu, L., Chen, R., Geirhofer, S., Sayana, K., Shi, Z., Zhou, Y. (2012) Downlink MIMO in LTE-advanced: SU-MIMO vs. MU-MIMO. IEEE Communications M 50(2): 140–147.

21. Lunden, J., Koivunen, V., Huttunen, A., Poor, H. V. (2009) Collaborative cyclostationary spectrum sensing for cognitive radio systems. IEEE T on Signal Processing 57(11): 4182–4195.

22. Mariani, A., Giorgetti, A., Chiani, M. (2011) SNR wall for energy detection with noise power estimation. In: Proceedings of IEEE International Conference on Communications (ICC), Kyoto, 5–9 June 2011.

23. Mitola, J., Maguire, G. Q. (1999) Cognitive radio: making software radios more personal. IEEE Personal Communications 6(4): 13–18.

24. Naraghi-Pour, M., Ikuma, T. (2010) Autocorrelation-based spectrum sensing for cognitive radios. IEEE T on Vehicular Technology 59(2): 718–733.

25. Nosratinia, A., Hunter, T. E., Hedayat, A. (2004) Cooperative communication in wireless networks. IEEE Communications M 42(10): 74–80.

26. Pawelczak, P., Nolan, K., Doyle, L., Oh, S. W., Cabric, D. (2011) Cognitive radio: Ten years of experimentation and development. IEEE Communications M 49(3): 90–100.

27. Perahia, E. (2008) IEEE 802.11n development: History, process, and technology. IEEE Communications M 46(7): 48–55.

28. Ramprashad, S. A., Papadopoulos, H. C., Benjebbour, A., Kishiyama, Y., Jindal, N., Caire, G. (2011) Cooperative cellular networks using multi-user MIMO: Trade-offs, overheads, and interference control across architectures. IEEE Communications M 49(5): 70–77.

29. Rusek, F., Persson, D., Lau, B. K., Larsson, E. G., Marzetta, T. L., Edfors, O., Tufvesson, F. (2013) Scaling up MIMO: Opportunities and challenges with very large arrays. IEEE Signal Processing M 30(1): 40–60.

30. Sahai, A., Hoven, N., Tandra, R. (2004) Some fundamental limits on cognitive radio. In: Proceedings of 42nd Allerton Conference on Communication, Control, and Computing, Monticello, 29 Sept-1 Oct 2004.

31. Spencer, Q. H., Peel, C. B., Swindlehurst, A. L., Haardt, M. (2004) An introduction to the multi-user MIMO downlink. IEEE Communications M 42(10): 60–67.

32. Sum, C. S., Harada, H., Kojima, F., Lu, L. (2013) An interference management protocol for multiple physical layers in IEEE 802.15.4g smart utility networks. IEEE Communications M 51(4): 84–91.

33. Wang, H., Yang, E. H., Zhao, Z., Zhang, W. (2009) Spectrum sensing in cognitive radio using goodness of fit testing. IEEE T on Wireless Communications 8(11): 5427–5430.

34. Wang, J., Ghosh, M., Challapali, K. (2011) Emerging cognitive radio applications: A survey. IEEE Communications M 49(3): 74–81.

35. Xu, W., Zhang, J., Zhang, P., Tellambura, C. (2012) Outage probability of decode-and-forward cognitive relay in presence of primary user's interference. IEEE Communications Letters 16(8): 1252–1255.

36. Yucek, T., Arslan, H. (2009) A survey of spectrum sensing algorithms for cognitive radio applications. IEEE Communications Surveys Tutorials 11(1): 116–130.

37. Zeng, Y., Liang, Y. C., Hoang, A. T., Zhang, R. (2010) A review on spectrum sensing for cognitive radio: Challenges and solutions. EURASIP J on Advances in Signal Processing.

38. Zeng, Y., Liang, Y. C. (2009) Eigenvalue-based spectrum sensing algorithms for cognitive radio. IEEE T on Communications 57(6): 1784–1793.

39. Zhang, G., Wang, X., Liang, Y. C., Liu, J. (2010) Fast and robust spectrum sensing via Kolmogorov-Smirnov test. IEEE T on Communications **58**(12): 3410–3416.
40. Zhang, Y. L., Zhang, Q. Y., Melodia, T. (2010) A frequency-domain entropy-based detector for robust spectrum sensing in cognitive radio networks. IEEE Communications Letters **14**(6): 533–535.
41. Zhao, Q., Sadler, B. M. (2007) A survey of dynamic spectrum access. IEEE Signal Processing M **24**(3): 79–89.

Chapter 2
Conventional Energy Detector

As mentioned before, the energy detector senses spectrum holes by determining whether the primary signal is absent or present in a given frequency slot. The energy detector typically operates without prior knowledge of the primary signal parameters. Its key parameters, including detection threshold, number of samples, and estimated noise power, determine the detection performance.

2.1 Binary Hypothesis Testing Problem

Depending on the idle state or busy state of the primary user, with the presence of the noise, the signal detection at the secondary user can be modeled as a binary hypothesis testing problem, given as

$$\text{Hypothesis 0 } (\mathcal{H}_0) : \text{ signal is absent}$$
$$\text{Hypothesis 1 } (\mathcal{H}_1) : \text{ signal is present.}$$

The transmitted signal of the primary user, denoted $\mathbf{s}$, is a complex signal. It has real component s_r and imaginary component s_i, i.e., $\mathbf{s} = s_r + js_i$.[1] If the received signal, $\mathbf{y}$, is sampled, the nth ($n = 1, 2, \cdots$) sample, $\mathbf{y}(n)$, can be given as [19, 24]

$$\mathbf{y}(n) = \begin{cases} \mathbf{w}(n) & : \mathcal{H}_0 \\ \mathbf{x}(n) + \mathbf{w}(n) & : \mathcal{H}_1 \end{cases} \quad (2.1)$$

---

[1] A complex number which has real and imaginary components z_r and z_i, respectively, is denoted as $\mathbf{z} = z_r + jz_i$. Unless otherwise specified, subscript "r" and "i" stand for real and imaginary component of a complex value, respectively.

S. Atapattu et al., *Energy Detection for Spectrum Sensing in Cognitive Radio*, SpringerBriefs in Computer Science, DOI 10.1007/978-1-4939-0494-5_2, © The Author(s) 2014

where $\mathbf{x}(n) = \mathbf{h}\mathbf{s}(n)$, $\mathbf{h}$ is channel gain (a complex value), and $\mathbf{w}(n) = w_r(n) + jw_i(n)$ is the noise sample which is assumed to be circularly symmetric complex Gaussian (CSCG) random variable with mean zero ($\mathbb{E}[\mathbf{w}(n)] = 0$) and variance $2\sigma_w^2$ ($\mathbb{V}\mathrm{ar}[\mathbf{w}(n)] = 2\sigma_w^2$), i.e., $\mathbf{w}(n) \sim \mathscr{C}\mathscr{N}(0, 2\sigma_w^2)$. Here $\mathbb{E}[\cdot]$ and $\mathbb{V}\mathrm{ar}[\cdot]$ are expectation and variance operations, respectively, and $\mathscr{C}\mathscr{N}(\cdot, \cdot)$ means a complex Gaussian distribution. Further, $w_r(n)$ and $w_i(n)$ are real-valued Gaussian random variables with mean zero and variance σ_w^2, i.e., $w_r(n), w_i(n) \sim \mathscr{N}(0, \sigma_w^2)$, where $\mathscr{N}(\cdot, \cdot)$ means a real Gaussian distribution. The channel gain denoted as $\mathbf{h} = h_r + jh_i$ is constant within each spectrum sensing period.

Equation (2.1) may be rewritten as

$$\mathbf{y}(n) = \theta\mathbf{x}(n) + \mathbf{w}(n) \tag{2.2}$$

where $\theta = 0$ for $\mathscr{H}_0$ and $\theta = 1$ for $\mathscr{H}_1$. In (2.1), perfect synchronization between the transmitter and the receiver is implicitly assumed. This assumption may not be valid for some practical situations, e.g., heavy-traffic in multi-user networks, in which primary users' signals arrive at the secondary receiver with a missed-matched sample durations n_0. Then, the signal model under $\mathscr{H}_1$ can be given as [36]

$$\mathscr{H}_1 : \ \mathbf{y}(n) = \begin{cases} \mathbf{w}(n) & : 1 \leq n \leq n_0 - 1 \\ \mathbf{x}(n) + \mathbf{w}(n) : & n_0 \leq n \leq N \end{cases} \tag{2.3}$$

where N is the total number of samples. This system model helps to analyze synchronizing uncertainty. For example, in high-traffic random access networks, when traffic patterns of transmitted signals are unknown to the receiver, the signal arrival time n_0 may be modeled as a random variable, e.g., uniformly distributed over the observation time.

In the literature, system model (2.1) is widely used, which is also focused on in the subsequent chapters. System model (2.3) is marginally investigated for spectrum sensing.

2.2   Energy Detection

The conventional energy detector measures the energy associated with the received signal over a specified time duration and bandwidth. The measured value is then compared with an appropriately selected threshold to determine the presence or the absence of the primary signal.

For theoretical analysis, two models of the conventional energy detector are considered in time-domain implementations:

- Analog energy detector (Fig. 2.1a) consists of a pre-filter followed by a square-law device and a finite time integrator [34]. The pre-filter limits the noise bandwidth and normalizes the noise variance. The output of the integrator is proportional to the energy of the received signal.

- Digital energy detector (Fig. 2.1b) consists of a low pass noise pre-filter that limits the noise and adjacent-bandwidth signals, an analog-to-digital converter (ADC) that converts continuous signals to discrete digital signal samples, and a square law device followed by an integrator.

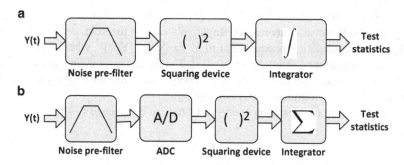

Fig. 2.1 The conventional energy detector: (**a**) analog and (**b**) digital

2.3   Test Statistic

The output of the analog or digital integrator (Fig. 2.1) is called decision (test) statistic. The test statistic is compared with the threshold to make the final decision on the presence/absence of the primary signal. However, the test statistic may not always be the integrator output, but a function that is monotonic with the integrator output [34].

When the Neyman-Pearson criterion is applied to the hypothesis problem in (2.1), the likelihood ratio for the binary hypothesis test given in (2.1) can be given as [33]

$$\Lambda_{LR} = \frac{f_{\mathbf{y}|\mathcal{H}_0}(x)}{f_{\mathbf{y}|\mathcal{H}_1}(x)} \tag{2.4}$$

where the probability density function (PDF) of the received signal $\mathbf{y}$ under hypotheses $\mathcal{H}$ is $f_{\mathbf{y}|\mathcal{H}}(x)$ where $\mathcal{H} \in \{\mathcal{H}_0, \mathcal{H}_1\}$. Then, the log-likelihood ratio (LLR) can be written as the form $a + b \sum_{n=1}^{N} |\mathbf{y}(n)|^2$ where N is the number of samples, and parameters a and b do not depend on the signal value $\mathbf{y}(n)$. Therefore, the LLR is proportional to $\sum_{n=1}^{N} |\mathbf{y}(n)|^2$ which is the test statistic of energy detector. This means, when the receiver knows only the received signal power, the energy detector is the optimal non-coherent detector for an unknown signal $\mathbf{s}(n)$ if $\mathbf{s}(n)$ is Gaussian, uncorrelated and independent with the uncorrelated background noise [32].

In digital implementation, after proper filtering, sampling, squaring and integration, the test statistic of energy detector is given by

$$\Lambda = \sum_{n=1}^{N} |\mathbf{y}(n)|^2 = \sum_{n=1}^{N} \left(e_r(n)^2 + e_i(n)^2 \right) \tag{2.5}$$

where $e_r(n) = \theta h_r s_r(n) - \theta h_i s_i(n) + w_r(n)$ and $e_i(n) = \theta h_r s_i(n) + \theta h_i s_r(n) + w_i(n)$. As the Parseval's theorem or Rayleigh's energy theorem, the test statistic of digital implementation is equivalent to $\Lambda = \sum_{k=1}^{N} |\mathbf{Y}(k)|^2$ where $\mathbf{Y}(k)$ is the frequency domain representation of $\mathbf{y}(n)$ [33]. The frequency domain representation is important for the spectrum sensing of energy detection under the orthogonal frequency division multiplexing (OFDM) system.

The test statistic of the analog energy detector can be given as [34]

$$\Lambda = \frac{1}{T} \int_{t-T}^{t} \mathbf{y}(t)^2 dt$$

where T is the time duration. A sample function with bandwidth W and time duration T can be described approximately by a set of samples $N \approx 2TW$, where TW is the time-bandwidth product [10]. Therefore, the analog test statistic can also be described by using digital one (2.5).

Moreover, the exact form of the test statistic may vary with applications. For example, in a heavy-traffic multi-user network, by using the hypothesis $\mathcal{H}_1$ (2.3), the test statistic may be defined as

$$\Lambda = \sum_{n=1}^{n_0-1} |\mathbf{y}(n)|^2 + \sum_{n=n_0}^{N} |\mathbf{y}(n)|^2$$

where there is only noise signal in the interval $[1, n_0 - 1]$. Moreover, for the analysis of parameter optimization or noise estimation error, Λ is usually normalized with respect to the sample number N and the noise variance $2\sigma_w^2$ as [20]

$$\Lambda = \frac{1}{2\sigma_w^2 N} \sum_{n=1}^{N} |\mathbf{y}(n)|^2. \tag{2.6}$$

The performance of energy detector (or of other detectors) is characterized by using following metrics, which have been introduced based on the test statistic under the binary hypothesis:

- False alarm probability (P_f): the probability of deciding the signal is present while $\mathcal{H}_0$ is true, i.e., $P_f = \mathrm{Pr}[\Lambda > \lambda | \mathcal{H}_0]$ where λ is the detection threshold, and $\mathrm{Pr}[\cdot]$ stands for an event probability. In the context of cognitive radio networks, a false alarm yields undetected spectrum holes. So a large P_f contributes to poor spectrum usage by secondary users.

- Missed-detection probability (P_{md}): the probability of deciding the signal is absent while $\mathcal{H}_1$ is true, i.e., $P_{md} = \Pr[\Lambda < \lambda | \mathcal{H}_1]$, which is equivalent to identifying a spectrum hole where there is none. Consequently, large P_{md} introduces unexpected interference to primary users.
- Detection probability (P_d): the probability of deciding the signal is present when $\mathcal{H}_1$ is true , i.e., $P_d = \Pr[\Lambda > \lambda | \mathcal{H}_1]$, and thus, $P_d = 1 - P_{md}$.

Both reliability and efficiency are expected from the spectrum sensing technique built into the cognitive radio, i.e., a higher P_d (or lower P_{md}) and lower P_f are preferred.

The statistical properties of Λ are necessary to characterize the performance of an energy detector. To get the statistical properties, signal and noise models are essential. While the noise components, $w_r(n)$ and $w_i(n)$, are often zero-mean Gaussian, different models for the signal to be detected are possible, as discussed below.

2.3.1 Signal Models

Based on the available knowledge of $\mathbf{s}(n)$, the receiver can adopt an appropriate model, which helps to analyze the distribution of the test statistic under $\mathcal{H}_1$. For example, three different models, **S1, S2** and **S3**, are popularly used in the literature, and are given as follows.

S1: For given channel gain $\mathbf{h}$, the signal to be detected, $\mathbf{y}(n)$, can be assumed as Gaussian with mean $\mathbb{E}[\mathbf{y}(n)] = \mathbb{E}[\mathbf{hs}(n) + \mathbf{w}(n)] = \mathbf{hs}(n)$ and variance $2\sigma_w^2$. This case may be modeled as an unknown deterministic signal. For the signal transmitted over a flat band-limited Gaussian noise channel, a basic mathematical model of the test statistic of an energy detector is given in [34]. The receive SNR can then be given as

$$\gamma_{S1} = \frac{|\mathbf{h}|^2 \frac{1}{N} \sum_{n=1}^{N} |\mathbf{s}(n)|^2}{2\sigma_w^2}. \tag{2.7}$$

S2: When the receiver has very limited knowledge of the transmitted signal (e.g., signal distribution), the signal sample may be considered as gaussian random variable, i.e., $\mathbf{s}(n) \sim \mathcal{CN}(0, 2\sigma_s^2)$, and then $\mathbf{y}(n) \sim \mathcal{CN}(0, 2(\sigma_w^2 + \sigma_s^2))$. The receive SNR can then be given as

$$\gamma_{S2} = \frac{|\mathbf{h}|^2 2\sigma_s^2}{2\sigma_w^2}. \tag{2.8}$$

S3: If the Gaussian assumption is removed from **S2** signal model, and signal sample is considered as random variable with mean zero and variance $2\sigma_s^2$, but with an unknown distribution, then $\mathbf{y}(n)$ has mean zero and variance $2(\sigma_w^2 + \sigma_s^2)$. The receive SNR can also be given as

$$\gamma_{S3} = \frac{|\mathbf{h}|^2 2\sigma_s^2}{2\sigma_w^2}. \tag{2.9}$$

For a sufficiently large number of samples, the signal variance can be written by using its sample variance as $2\sigma_s^2 \approx \frac{1}{N}\sum_{n=1}^{N}|\mathbf{s}(n)|^2 - |\frac{1}{N}\sum_{n=1}^{N}\mathbf{s}(n)|^2$. If the sample mean goes to zero, i.e., when $\frac{1}{N}\sum_{n=1}^{N}\mathbf{s}(n) \to 0$, we have $2\sigma_s^2 \approx \frac{1}{N}\sum_{n=1}^{N}|\mathbf{s}(n)|^2$, and thus, all the receive SNRs given in (2.7)–(2.9) under different signal models have the same expression. In this case, the instantaneous SNR is denoted as γ.

2.3.2 Distribution of Test Statistics

The exact distributions of test statistics given in (2.5) for different signal models are analyzed in the following under both hypotheses, $\mathcal{H}_0$ and $\mathcal{H}_1$. The PDFs of Λ under hypotheses $\mathcal{H}_0$ and $\mathcal{H}_1$ are denoted as $f_{\Lambda|\mathcal{H}_0}(x)$ and $f_{\Lambda|\mathcal{H}_1}(x)$, respectively.

2.3.2.1 Under $\mathcal{H}_0$

In this case, $e_r(n) = w_r(n)$ and $e_i(n) = w_i(n)$, and $e_r(n)$ and $e_i(n)$ follow $\mathcal{N}(0, \sigma_w^2)$. Thus, Λ is a sum of $2N$ squares of independent $\mathcal{N}(0, \sigma_w^2)$ random variables, and it follows central chi-square distribution given as [23]

$$f_{\Lambda|\mathcal{H}_0}(x) = \frac{x^{N-1}e^{-\frac{x}{2\sigma_w^2}}}{\left(2\sigma_w^2\right)^N \Gamma(N)} \tag{2.10}$$

where $\Gamma(n) = \int_0^\infty t^{n-1}e^{-t}dt$ is the gamma function [12]. Thus, the false-alarm probability can be derived, $P_f = \Pr[\Lambda > \lambda|\mathcal{H}_0]$, by using (2.10) as

$$P_f = \frac{\Gamma(N, \frac{\lambda}{2\sigma_w^2})}{\Gamma(N)} \tag{2.11}$$

where $\Gamma(n, x) = \int_x^\infty t^{n-1}e^{-t}dt$ is the upper incomplete gamma function [12].

2.3.2.2 Under $\mathcal{H}_1$

In this case, the distribution of Λ, $f_{\Lambda|\mathcal{H}_1}(x)$, has two different distributions under two signal models, **S1** and **S2**, for a given channel. However, the distribution of Λ under **S3** cannot be derived.

For **S1**, $e_r(n)$ follows $\mathcal{N}(h_r s_r(n) - h_i s_i(n), \sigma_w^2)$, and $e_i(n)$ follows $\mathcal{N}(h_r s_i(n) + h_i s_r(n), \sigma_w^2)$. Since Λ is a sum of $2N$ squares of independent and non-identically distributed Gaussian random variables with non-zero mean, Λ follows non-central chi-square distribution given as [23]

$$f_{\Lambda|\mathscr{H}_1}(x) = \frac{\left(\frac{x}{\sigma_w^2}\right)^{\frac{N-1}{2}} e^{-\frac{1}{2}\left(\frac{x}{\sigma_w^2}+\mu\right)}}{2\sigma_w^2 \mu^{\frac{N-1}{2}}} I_{N-1}\left(\sqrt{\mu \frac{x}{\sigma_w^2}}\right), \quad 0 \le x \le \infty, \qquad (2.12)$$

where $I_\nu(\cdot)$ is the modified Bessel function of the first kind of order ν,

$$\mu = \sum_{n=1}^{N}\left[\frac{(h_r s_r(n) - h_i s_i(n))^2}{\sigma_w^2} + \frac{(h_r s_i(n) + h_i s_r(n))^2}{\sigma_w^2}\right] = 2N\gamma_{S1}$$

which is the non-centrality parameter, and γ_{S1} is given in (2.7). Thus, the detection probability, $P_d = \Pr(\Lambda > \lambda|\mathscr{H}_1)$, can be derived for **S1** by using (2.12) as

$$P_{d,S1} = Q_N\left(\sqrt{2N\gamma_{S1}}, \frac{\sqrt{\lambda}}{\sigma_w}\right) \qquad (2.13)$$

where $Q_N(a,b) = \int_b^\infty x \left(\frac{x}{a}\right)^{N-1} e^{-\frac{x^2+a^2}{2}} I_{N-1}(ax)dx$ is the generalized Marcum-Q function [22]. This signal model is widely used in the performance analysis of an energy detector in terms of the average detection probability [1–4, 9, 10, 14–17].

For **S2**, $e_r(n)$ and $e_i(n)$ follow $\mathcal{N}(0, (1 + \gamma_{S2})\sigma_w^2)$ where γ_{S2} is given in (2.8). Since Λ is a sum of $2N$ squares of independent and identically distributed (i.i.d.) Gaussian random variables with zero mean, Λ follows central chi-square distribution which is given as

$$f_{\Lambda|\mathscr{H}_1}(x) = \frac{x^{N-1} e^{-\frac{x}{2(1+\gamma_{S2})\sigma_w^2}}}{\left(2(1 + \gamma_{S2})\sigma_w^2\right)^N \Gamma(N)}. \qquad (2.14)$$

The exact detection probability can be derived for **S2** by using (2.14) as

$$P_{d,S2} = \frac{\Gamma\left(N, \frac{\lambda}{2\sigma_w^2(1+\gamma_{S2})}\right)}{\Gamma(N)}. \qquad (2.15)$$

This model is used in [6, 27].

For **S3**, $e_r(n)$ and $e_i(n)$ have unknown distributions, and the exact $f_{\Lambda|\mathscr{H}_1}(x)$ cannot be derived, and a central or non-central chi-square distribution may not work. However, $f_{\Lambda|\mathscr{H}_1}(x)$ can be derived approximately by using the central limit theorem (CLT).

2.3.3 CLT Approach

The CLT suggests that the sum of N i.i.d. random variables with finite mean and variance approaches a normal distribution when N is large enough. Using the CLT, the distribution of the test statistic (2.5) can be accurately approximated with a normal distribution for a sufficiently large number of samples as

$$\Lambda \sim \mathcal{N} \left(\sum_{n=1}^{N} \mathbb{E}[|\mathbf{y}(n)|^2], \sum_{n=1}^{N} \mathbb{V}\text{ar}[|\mathbf{y}(n)|^2] \right).$$

The mean and variance for different signal models are given as follows:

$$\mathbb{E}[|\mathbf{y}(n)|^2] = \begin{cases} 2\sigma_w^2 & : \mathcal{H}_0 \\ 2\sigma_w^2 + |\mathbf{h}|^2 |\mathbf{s}(n)|^2 & : \mathbf{S1} \\ 2\sigma_w^2 + |\mathbf{h}|^2 (2\sigma_s^2) & : \mathbf{S2} \\ 2\sigma_w^2 + |\mathbf{h}|^2 (2\sigma_s^2) & : \mathbf{S3}. \end{cases} \tag{2.16}$$

$$\mathbb{V}\text{ar}[|\mathbf{y}(n)|^2] = \begin{cases} (2\sigma_w^2)^2 & : \mathcal{H}_0 \\ 4\sigma_w^2(\sigma_w^2 + |\mathbf{h}|^2 |\mathbf{s}(n)|^2) & : \mathbf{S1} \\ 4(\sigma_w^2 + |\mathbf{h}|^2 \sigma_s^2)^2 & : \mathbf{S2} \\ (2\sigma_w^2)^2 + 2|\mathbf{h}|^2 (2\sigma_w^2)(2\sigma_s^2) + |\mathbf{h}|^4 (\mathbb{E}[|\mathbf{s}(n)|^4] - 4\sigma_s^4) & : \mathbf{S3}. \end{cases} \tag{2.17}$$

If $\mathbf{s}(n)$ of **S3** is complex phase-shift keying (PSK) signal, $\mathbb{E}[|\mathbf{s}(n)|^4] = 4\sigma_s^4$, and thus the variance can be evaluated as $\mathbb{V}\text{ar}[|\mathbf{y}(n)|^2] = (2\sigma_w^2)^2 + 2|\mathbf{h}|^2 (2\sigma_w^2)(2\sigma_s^2)$. This will be used in the following sections. Therefore, the distribution of Λ can be given as

$$\Lambda \sim \begin{cases} \mathcal{N}\left(N(2\sigma_w^2), N(2\sigma_w^2)^2\right) & : \mathcal{H}_0 \\ \mathcal{N}\left(N(2\sigma_w^2)(1+\gamma), N(2\sigma_w^2)^2(1+2\gamma)\right) & : \mathbf{S1}, \mathbf{S3} \text{ (complex-PSK)} \\ \mathcal{N}\left(N(2\sigma_w^2)(1+\gamma), N(2\sigma_w^2)^2(1+\gamma)^2\right) & : \mathbf{S2}. \end{cases} \tag{2.18}$$

By using each mean and variance in (2.18), an approximated false-alarm probability is

$$P_f \approx Q\left(\frac{\lambda - N(2\sigma_w^2)}{\sqrt{N}(2\sigma_w^2)}\right) \tag{2.19}$$

where $Q(x) = \frac{1}{\sqrt{2\pi}} \int_x^\infty e^{-\frac{u^2}{2}} du$ is the Gaussian-Q function. Similarly, approximated detection probabilities are

$$P_{d,S1} \approx Q\left(\frac{\lambda - N(2\sigma_w^2)(1+\gamma)}{\sqrt{N(1+2\gamma)}(2\sigma_w^2)}\right), \tag{2.20}$$

$$P_{d,S2} \approx Q\left(\frac{\lambda - N(2\sigma_w^2)(1+\gamma)}{\sqrt{N}(1+\gamma)(2\sigma_w^2)}\right). \tag{2.21}$$

Note that $P_{d,S3}$ has the same expression as $P_{d,S1}$.

Figure 2.2 shows the exact P_f and P_d of the test statistic for **S1** and **S2**.[2] Both P_f and P_d increase significantly when the number of samples increases from $N = 20$ to $N = 60$. For the same SNR (i.e., $\gamma_{S1} = \gamma_{S2}$), **S2** outperforms **S1**. However, this may not be a rigorous comparison because γ_{S1} and γ_{S2} have two different definitions. Figure 2.2 also shows the CLT approximations of P_f and P_d. The exact curves (solid-line) match well with the CLT approximations (dashed-line) when $N = 60$, while they have a close match when $N = 20$. This confirms the validity of the CLT approximation for the distribution of the test statistic for a sufficiently large number of samples.

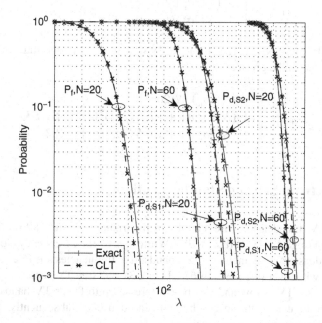

Fig. 2.2 The exact and approximated (CLT) cumulative distribution function (CDF) of the test statistic for **S1** and **S2** with $2\sigma_w^2 = 1$ and $\gamma = 5\,\mathrm{dB}$

---

[2] P_f and P_d are also the complementary CDFs of the test statistic under $\mathcal{H}_0$ and $\mathcal{H}_1$, respectively.

The Gaussian approximation is commonly used to optimize system parameters (such as detection threshold or sensing time), because this model often gives a more convenient cost function [18, 19, 24]. As well, the inverse of the Gaussian-$Q(\cdot)$ function can be readily derived in closed form. Moreover, well-known mathematically-tractable models based on the Gaussian approximation have further been introduced in the literature in order to simplify the performance analysis (e.g., Edell Model, Berkeley Model and Torrieri Model) [7,21].

2.3.4 Low-SNR and High-SNR Models

At low SNR, a reliable detection is possible with a large N. Thus, the CLT gives good approximations for P_f and P_d. As SNR$\ll 1$ (e.g., -20 dB), $1 + 2\gamma \approx 1$ or $1 + \gamma \approx 1$, and therefore, the signal has little impact on the variance of the test statistic given in (2.18). Thus, a low-SNR approximation can be given for any of the three signal models (**S1**, **S2**, and **S3**) as

$$\Lambda_{\text{low}} \sim \begin{cases} \mathcal{N}\left(N(2\sigma_w^2), N(2\sigma_w^2)^2\right) & : \mathcal{H}_0 \\ \mathcal{N}\left(N(2\sigma_w^2)(1+\gamma), N(2\sigma_w^2)^2\right) & : \mathcal{H}_1. \end{cases} \tag{2.22}$$

At high SNR, i.e., SNR$\gg 1$, $1 + 2\gamma \approx 2\gamma$ or $1 + \gamma \approx \gamma$, and thus a high-SNR approximation can be given as

$$\Lambda_{\text{high}} \sim \begin{cases} \mathcal{N}\left(N(2\sigma_w^2), N(2\sigma_w^2)^2\right) & : \mathcal{H}_0 \\ \mathcal{N}\left(N(2\sigma_w^2)(1+\gamma), N(2\sigma_w^2)^2(2\gamma)\right) & : \textbf{S1, S3} \\ \mathcal{N}\left(N(2\sigma_w^2)(1+\gamma), N(2\sigma_w^2)^2\gamma^2\right) & : \textbf{S2}. \end{cases} \tag{2.23}$$

2.4 Spectrum Sensing Standardization

Different TV broadcasters use chunk of the radio spectrum allowed for TV broadcasting (e.g., 54–806 MHz in US). White spaces (i.e., frequency slots unused by TV broadcasters) may include guard bands, free frequencies due to analog TV to digital TV switchover (e.g., 698–806 MHz in US), and free TV bands created when traffic in digital TV is low and can be compressed into fewer TV bands. The US FCC allows to use white spaces by unlicensed users. Subsequently, following standardization efforts have materialized:

- The IEEE 802.22 standard for TV white spaces has been released with medium access control and physical layer specifications for WRAN.
- The ECMA 392 includes specification for personal/portable wireless devices operating in TV bands [11],

- The IEEE SCC41 develops supporting standards for radio and dynamic spectrum management [13].
- The IEEE 802.11af is for Wi-Fi on the TV white spaces using cognitive radio technology [30].

2.4.1 IEEE 802.22 Standard

Among the above standards efforts, IEEE 802.22 WRAN brings broadband access not only to Wi-Fi devices but also to general mobile networks (e.g., micro-, pico- or femto-cells), allowing the use of the cognitive radio technique on a non-interfering basis [8, 28, 29]. Since the IEEE 802.22 WRAN does not prescribe a specific spectrum sensing technique, designers are free to select any detection technique. Therefore, energy detection is one of the most obvious choices. In implementation of energy detection, the specifications that have to be considered carefully are given below:

- The IEEE 802.22 WRAN limits both false alarm (which indicates the level of undetected spectrum holes) and missed-detection (which indicates the level of unexpected interference to primary users) probabilities to 10%.
- While false alarm and missed-detection probabilities reflect the overall efficiency and reliability of the cognitive network, the 10% requirement should be met even under very low SNR conditions, such as $-20\,$dB SNR with a signal power of $-116\,$dBm and a noise floor of $-96\,$dBm [29].
- While energy detector performs well at moderate and high SNRs, it performs poorly at a low SNR. Although increasing the sensing time is an obvious option for improving the sensing performance, IEEE 802.22 limits the maximal detection latency to 2 s which includes sensing time and subsequent processing time. This maximal time limit is critical at low-SNR spectrum sensing.

Thus, energy detector parameters must be designed carefully based on spectrum sensing specifications.

2.5 Design Parameters

The main design parameters of the energy detector are the number of samples and threshold. Although the performance of the energy detector depends on SNR and noise variance as well, designers have very limited control over them because these parameters depend on the behavior of the wireless channel.

2.5.1 Threshold

A pre-defined threshold λ is required to decide whether the target signal is absent or present. This threshold determines all performance metrics, P_d, P_f and P_{md}. Since it varies from 0 to ∞, selection of operating threshold is important. The operating threshold thus can be determined based on the target value of the performance metric of interest.

When the threshold increases (or decreases), both P_f and P_d decrease (or increase). For known N and σ_w, the common practice of setting the threshold is based on a constant false alarm probability P_f, e.g., $P_f \leq 0.1$. The selected threshold based on P_f can be given by using (2.19) as

$$\lambda_f^* = \left(Q^{-1}(P_f) + \sqrt{N} \right) \sqrt{N} 2\sigma_w^2. \tag{2.24}$$

However, this threshold may not guarantee that the energy detector achieves the target detection probability (e.g., 0.9 specified in the IEEE 802.22 WRAN). Thus, threshold selection can be viewed as an optimization problem to balance the two conflicting objectives (i.e., maximize P_d while minimizing P_f).

2.5.2 Number of Samples

The number of samples (N) is also an important design parameter to achieve the requirements on detection and false alarm probabilities. For given false alarm probability P_f and detection probability P_d, the minimum required number of samples can be given as a function of SNR. By eliminating λ from both P_f in (2.19) and P_d in (2.20) (here signal model **S1** is used as an example), N can be given as

$$N = \left[Q^{-1}(P_f) - Q^{-1}(P_d)\sqrt{2\gamma + 1} \right]^2 \gamma^{-2} \tag{2.25}$$

which is not a function of the threshold. Due to the monotonically decreasing property of function $Q^{-1}(\cdot)$, it can be seen that the signal can be detected even in very low SNR region by increasing N when the noise power is perfectly known. Further, the approximate required number of samples to achieve a performance target on false alarm and detection probabilities is in the order of $\mathcal{O}(\gamma^{-2})$, i.e., energy detector requires more samples at very low SNR [5]. Since $N \approx \tau f_s$ where τ is the sensing time and f_s is the sampling frequency, the sensing time increases as N increases. This is a main drawback in spectrum sensing at low SNR because of the limitation on the maximal allowable sensing time (e.g., the IEEE 802.22 specifies that the sensing time should be less than 2 s). Therefore, the selection of N is also an optimization problem.

2.6   Noise Effect

Threshold selection depends on the noise power. A proper threshold selection, as given in (2.24), is possible only if noise power is accurately known at the receiver. As in (2.25), when the SNR is small ($\gamma \to 0$), the number of samples increases ($N \to \infty$). This means that the given P_f and P_d can be achieved even with small SNR by using a large number of samples. This is possible only if noise power is accurately known [5, 25, 26, 31, 32].

The accurate noise-power estimation is not always possible. Since noise may include the effects of nearby interference from other transmissions, weak signals, temperature changes, and filtering effect, the additive white Gaussian noise (AWGN) properties (e.g., wideband noise with a constant spectral density) of the resultant noise may be lost, which affects the noise power estimation [35]. The estimation error is referred to as noise uncertainty, an error that can seriously degrade the energy detector performance. With noise uncertainty, the estimated noise power is assumed to be in an interval $[\frac{1}{\rho}\sigma_w^2, \rho\sigma_w^2]$ where ρ (> 1) is the parameter that quantifies the noise uncertainty [32]. By using the low-SNR approximation $2\gamma + 1 \approx 1$ and the noise uncertainty effect, the required number of samples for the conventional energy detector to achieve given P_f and P_d can be given as

$$N \approx \frac{(Q^{-1}(P_f) - Q^{-1}(P_d))^2}{(\gamma - (\rho - \frac{1}{\rho}))^2}. \tag{2.26}$$

This indicates that an infinitely large number of samples are necessary to achieve the target false-alarm and detection probabilities when $\gamma \to (\rho - \frac{1}{\rho})$. A practical energy detector cannot be implemented at this SNR level, referred to as *SNR wall phenomenon*.

The test statistic of a practical energy detector which uses estimated noise power (ENP) is thus defined as [20]

$$\Lambda_{\text{ENP}} = \frac{1}{2\hat{\sigma}_w^2 N} \sum_{n=1}^{N} |\mathbf{y}(n)|^2 \tag{2.27}$$

where $2\hat{\sigma}_w^2$ is the estimated noise variance. In this case, the SNR wall is derived as

$$\gamma_{\text{min}} = \frac{1 - Q^{-1}(P_d)\sqrt{\phi}}{1 - Q^{-1}(P_f)\sqrt{\phi}} - 1 \tag{2.28}$$

where $\phi = \text{Var}\left(\frac{\hat{\sigma}_w^2}{\sigma_w^2}\right)$. In practice, noise can be estimated by using noise-only samples, which is similar to hypothesis under $\mathcal{H}_0$. Denote M as the number of noise-only samples. Then, ϕ can also be given as $\phi = \sqrt{\frac{N+M}{NM}}$ [20].

The energy detector needs an infinitely large number of samples if $\gamma \to \gamma_{\min}$. This implies that target false-alarm and detection probabilities cannot be achieved even with a large number of samples if $\gamma < \gamma_{\min}$. Figure 2.3 shows $\gamma_{\min}$ variation with N for both ideal and ENP energy detectors for $P_f = 0.1$ and $P_d = 0.9$. For ideal energy detector, $\gamma_{\min}$ decreases as N increases, which means that target false-alarm and detection probabilities can be achieved at very low SNR by increasing N. But for ENP energy detector with $M = 100$, if SNR $\gamma < -5.3\,\mathrm{dB}$ the desirable performance, i.e., $P_f = 0.1$ and $P_d = 0.9$, cannot be achieved at any N. However, $\gamma_{\min}$ can be decreased by increasing M, e.g., $\gamma_{\min} \to -10.7\,\mathrm{dB}$ at $M = 1000$ [20].

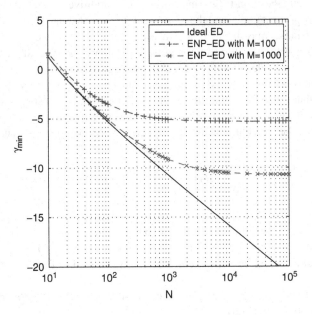

Fig. 2.3 Variation of $\gamma_{\min}$ with N of ideal and ENP energy detectors (*ED*)

References

1. Atapattu, S., Tellambura, C., Jiang, H. (2009) Energy detection of primary signals over $\eta - \mu$ fading channels. In: Proceedings of International Conference Industrial and Information Systems (ICIIS), Kandy, 28–31 Dec 2009.
2. Atapattu, S., Tellambura, C., Jiang, H. (2009) Relay based cooperative spectrum sensing in cognitive radio networks. In: Proceedings of IEEE Global Telecommunications Conference (GLOBECOM), Hawaii, 30 Nov- 4 Dec 2009.
3. Atapattu, S., Tellambura, C., Jiang, H. (2010) Analysis of area under the ROC curve of energy detection. IEEE T on Wireless Communications **9**(3): 1216–1225.

4. Atapattu, S., Tellambura, C., Jiang, H. (2010) Performance of an energy detector over channels with both multipath fading and shadowing. IEEE T on Wireless Communications 9(12): 3662–3670.

5. Cabric, D., Tkachenko, A., Brodersen, R. W. (2006) Experimental study of spectrum sensing based on energy detection and network cooperation. In: Proceedings of International Workshop on Technology and Policy for Accessing Spectrum (TAPAS), Boston, 5 Aug 2006.

6. Chen, Y. (2010) Improved energy detector for random signals in Gaussian noise. IEEE T on Wireless Communications 9(2): 558–563.

7. Ciftci, S., Torlak, M. (2008) A comparison of energy detectability models for spectrum sensing. In: Proceedings of IEEE Global Telecommunications Conference (GLOBECOM), New Orleans, 30 Nov-4 Dec 2008.

8. Cordeiro, C., Challapali, K., Birru, D., Shankar, S. N. (2006) IEEE 802.22: An introduction to the first wireless standard based on cognitive radios. J of Communications (JCM) 1(1): 38–47.

9. Digham, F. F., Alouini, M. S., Simon, M. K. (2003) On the energy detection of unknown signals over fading channels. In: Proceedings of IEEE International Conference on Communications (ICC), Anchorage, 11–15 May 2003.

10. Digham, F. F., Alouini, M. S., Simon, M. K. (2007) On the energy detection of unknown signals over fading channels. IEEE T on Communications 55(1): 21–24.

11. ECMA-International (2012) MAC and PHY for operation in TV white space. http://www.ecma-international.org/publications/files/ECMA-ST/ECMA-392.pdf.

12. Gradshteyn, I. S., Ryzhik, I. M. (2000) Table of Integrals, Series, and Products, 6th edn, Academic Press, Inc.

13. Granelli, F., Pawelczak, P., Prasad, R. V., Subbalakshmi, K. P., Chandramouli, R., Hoffmeyer, J. A. and Berger, H. S. (2010) Standardization and research in cognitive and dynamic spectrum access networks: IEEE SCC41 efforts and other activities. IEEE Communications M 48(1): 71–79.

14. Herath, S. P., Rajatheva, N. (2008) Analysis of equal gain combining in energy detection for cognitive radio over Nakagami channels. In: Proceedings of IEEE Global Telecommunications Conference (GLOBECOM), New Orleans, 30 Nov-4 Dec 2008.

15. Herath, S. P., Rajatheva, N., Tellambura, C. (2009) On the energy detection of unknown deterministic signal over Nakagami channels with selection combining. In: Canadian Conference on Electrical and Computing Engineering (CCECE), Newfoundland, 3–6 May 2009.

16. Herath, S. P., Rajatheva, N., Tellambura, C. (2009) Unified approach for energy detection of unknown deterministic signal in cognitive radio over fading channels. In: Proceedings of IEEE International Conference on Communications (ICC) Workshops, Dresden, 14–18 June 2009.

17. Kostylev, V. I. (2002) Energy detection of a signal with random amplitude. In: Proceedings of IEEE International Conference on Communications (ICC), New York City, 28 Apr-2 May 2002.

18. Liang, Y. C., Zeng, Y., Peh, E. C. Y., Hoang, A. T. (2007) Sensing-throughput tradeoff for cognitive radio networks. In: Proceedings of IEEE International Conference on Communications (ICC), Glasgow, 24–28 June 2007.

19. Liang, Y. C., Zeng, Y., Peh, E. C. Y., Hoang, A. T. (2008) Sensing-throughput tradeoff for cognitive radio networks. IEEE T on Wireless Communications 7(4): 1326–1337.

20. Mariani, A., Giorgetti, A., Chiani, M. (2011) Effects of noise power estimation on energy detection for cognitive radio applications. IEEE T on Communications 59(12): 3410–3420.

21. Mills, R., Prescott, G. (1996) A comparison of various radiometer detection models. IEEE T on Aerospace and Electronic Systems 32(1): 467–473.

22. Nuttall, A. H. (1974) Some integrals involving the Q_M-function. Naval underwater Systems Center (NUSC) technical report.

23. Papoulis, A., Pillai, S. U. (2002) Probability, Random Variables and Stochastic Processes, McGraw-Hill Companies, Inc.

24. Quan, Z., Cui, S., Sayed, A. H., Poor, H. V. (2009) Optimal multiband joint detection for spectrum sensing in cognitive radio networks. IEEE T on Signal Processing 57(3); 1128–1140.

25. Sahai, A., Hoven, N., Tandra, R. (2004) Some fundamental limits on cognitive radio. In: Proceedings of 42nd Allerton Conference on Communication, Control, and Computing, Monticello, 29 Sept-1 Oct 2004.
26. Sahai, A., Tandra, R., Mishra, S. M., Hoven, N. (2006) Fundamental design tradeoffs in cognitive radio systems. In: Proceedings of International Workshop on Technology and Policy for Accessing Spectrum (TAPAS), Boston, 5 Aug 2006.
27. Salt, J. E., Nguyen, H. H. (2008) Performance prediction for energy detection of unknown signals. IEEE T on Vehicular Technology **57**(6), 3900–3904.
28. Shellhammer, S. J. (2008) Spectrum sensing in IEEE 802.22. In: 1st IAPR Workshop on Cognitive Information Processing, Santorini (Thera), 9–10 June 2008.
29. Stevenson, C., Chouinard, G., Lei, Z., Hu, W., Shellhammer, S. J., Caldwell, W. (2009) IEEE 802.22: The first cognitive radio wireless regional area network standard. IEEE Communications M **47**(1): 130–138.
30. Sum, C. S., Harada, H., Kojima, F., Lan, Z., Funada, R. (2011) Smart utility networks in TV white space. IEEE Communications M **49**(7), 132–139.
31. Tandra, R., Sahai, A. (2005) Fundamental limits on detection in low SNR under noise uncertainty. In: International Conference on Wireless Networks, Communications and Mobile Computing (WCNM), Wuhan, 13–16 June 2005
32. Tandra, R., Sahai, A. (2008) SNR walls for signal detection. IEEE J on Selected Topics in Signal Processing **2**(1): 4–17.
33. Trees, H. L. V. (2001) Detection, Estimation, and Modulation Theory, Part I, Wiley-Interscience.
34. Urkowitz, H. (1967) Energy detection of unknown deterministic signals. Proceedings of the IEEE **55**(4): 523–531.
35. Vijayandran, L., Dharmawansa, P., Ekman, T., Tellambura, C. (2012) Analysis of aggregate interference and primary system performance in finite area cognitive radio networks. IEEE T on Communications **60**(7): 1811–1822.
36. Wu, J. Y., Wang, C. H., Wang, T. Y. (2011) Performance analysis of energy detection based spectrum sensing with unknown primary signal arrival time. IEEE T on Communications **59**(7): 1779–1784.

Chapter 3
Alternative Forms of Energy Detectors

Despite the benefits of low complexity and simple structure, the energy detector has several drawbacks.

- Performance degrades for low SNR. To avoid this, a longer time duration (or a larger number of samples) to gather enough samples of the received signal helps.
- Reliable detection for a large SNR range may not be possible. Moreover, inaccurate noise estimation due to rapid noise-power fluctuation and unknown background interference causes the noise uncertainty.
- It is optimal for independent Gaussian noise samples only. When noise samples are dependent or non-Gaussian due to unknown background interference, the test statistic (2.5) is not optimal.

To circumvent these drawbacks, improved energy detectors have been introduced, and several are briefly explained in the following.

3.1 Probability-Based Weighted Energy Detector

The frame format of the secondary network consists of sensing and data blocks. In the conventional energy detection, the signal samples of primary user are assumed to occur at the beginning of the sensing block, and hence each received signal sample is weighted equally, as given in (2.1). However, in practice, the primary samples may occur at anytime within the sensing block. Moreover, the Neyman-Pearson criterion shows that the energy detector maximizes the detection probability for a given false alarm probability when samples are weighted according to the primary user's appearance in the sensing block. This is the key idea of the probability-based energy detector [4, 11, 12].

Since the primary samples can occur at the mth ($1 \leq m \leq N$) sample of the current sensing block, each sample can be considered as different hypotheses. There are N different cases corresponding to the appearance of the primary user at N

S. Atapattu et al., *Energy Detection for Spectrum Sensing in Cognitive Radio*,
SpringerBriefs in Computer Science, DOI 10.1007/978-1-4939-0494-5_3,
© The Author(s) 2014

different samples within the sensing block. Therefore, the mth hypothesis, $\mathcal{H}_m$, is a sub-hypothesis of a composite hypothesis testing. The received signal under $\mathcal{H}_m$ is given [similar to (2.1)] as

$$
\mathbf{y}_{\mathcal{H}_m}(n) = \begin{cases} \mathbf{w}(n) : & 1 \leq n \leq m-1 \\ \mathbf{x}(n) + \mathbf{w}(n) : & m \leq n \leq N. \end{cases} \tag{3.1}
$$

If the received signal vector is $\mathbf{Y}$, the likelihood ratio for the binary hypothesis testing of idle and busy states can be given as

$$
\frac{f_{\mathbf{Y}|\mathcal{H}_m}(\mathbf{y}(n))}{f_{\mathbf{Y}|\mathcal{H}_0}(\mathbf{y}(n))} = \left(\frac{1}{1+\gamma}\right)^{N-m+1} e^{\frac{1}{1+\gamma}\sum_{n=m}^{N}|\mathbf{y}(n)|^2} \tag{3.2}
$$

where $\gamma = \frac{|\mathbf{h}|^2 2\sigma_s^2}{2\sigma_w^2}$ which corresponds to the signal model **S2**. The test statistic of (3.2) depends only on the observation of $\sum_{n=m}^{N}|\mathbf{y}(n)|^2$, which is a simple energy detector within the observation samples in $m \leq n \leq N$. Since there is no signal in the sample interval $1 \leq n \leq m-1$ and there is signal in the sample interval $m \leq n \leq N$, a weighting factor, $w_{\mathcal{H}_m,n}$, can be introduced to distinguish the two cases as

$$
w_{\mathcal{H}_m,n} = \begin{cases} 0 : 1 \leq n \leq m-1 \\ 1 : & m \leq n \leq N. \end{cases} \tag{3.3}
$$

Since the primary signal randomly appears in the sensing block, idle and busy durations of the primary user are random. Thus, m is also a discrete random number. As m is unknown, weighting factors in (3.3) cannot be applied directly. Therefore, by using the randomness of m, weighting factors can be designed based on the probability mass function of m, $p(m)$. Then, the test statistic of the probability-based weighted energy detector is given as

$$
\Lambda = \sum_{n=1}^{N} w_n |\mathbf{y}(n)|^2 \tag{3.4}
$$

where $w_i = \frac{\sum_{m=1}^{i} p(m)}{\sum_{m=1}^{N} p(m)}$.

The distributions of idle and busy durations of the primary user are given in the literature [32]. Assuming idle duration of the licensed channel is exponentially distributed, the probability-based energy detector is extensively investigated in [12]. This detector is shown to achieve nearly optimal performance with relatively low complexity. Since $p(m)$ is needed to compute the weighting factors, it can be estimated by observing statistics of licensed channel occupancy. This estimation however requires training techniques, increasing the overhead and complexity.

3.2 Double Threshold Energy Detector

The conventional energy detector has single threshold value as shown in Fig. 3.1a. The precise threshold, λ^*, which achieves the target probabilities of false-alarm and detection, may not be accurately computed by using threshold selection techniques because of inaccurate parameter estimations, e.g., noise estimation error. Thus, the selected threshold can be given as $\lambda_s = \lambda^* \pm \epsilon$ where ϵ quantifies the threshold uncertainty. This threshold uncertainty may significantly impact the expected detection performance. If the threshold uncertainty range can be reduced, reliable decision can be expected from the energy detector. This idea motivates the double-threshold energy detector which has two threshold values, λ_1 and λ_2 where $\lambda_1 < \lambda_2$ as shown in Fig. 3.1b [19, 23, 33].

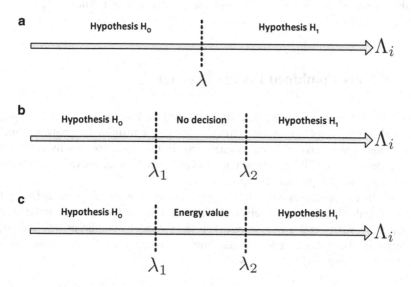

Fig. 3.1 Threshold of an energy detector: (**a**) single-threshold λ, (**b**) double-threshold λ_1 and λ_2 where there is no decision taken in $\lambda_1 < \Lambda < \lambda_2$, and (**c**) double-threshold λ_1 and λ_2 where the energy value is used when $\lambda_1 < \Lambda < \lambda_2$

The double-threshold energy detector's decision is $\mathcal{H}_0$ or $\mathcal{H}_1$ if $\Lambda \leq \lambda_1$ or $\Lambda \geq \lambda_2$, respectively, and there is no decision if $\lambda_1 < \Lambda < \lambda_2$. The double threshold energy detector is usually used in decision-fusion based cooperative spectrum sensing networks, where a number of cooperative nodes make their individual decisions and forward to the fusion center. If the test statistic of the ith cooperative node is Λ_i, the local decision of the ith cooperative node, D_i, is given as

$$D_i = \begin{cases} \mathcal{H}_0; & \Lambda_i \leq \lambda_1, \\ \text{No decision} \; ; & \lambda_1 < \Lambda_i < \lambda_2, \\ \mathcal{H}_1; & \Lambda_i \geq \lambda_2. \end{cases} \qquad (3.5)$$

While cooperative nodes with reliable decision, i.e., $D_i = \mathcal{H}_0$ or $D_i = \mathcal{H}_1$, report their individual decisions to the fusion center, other nodes do not. This method not only minimizes the total network energy consumption at the detection stage, but also reduces the network traffic by avoiding reports from cooperative nodes with unreliable decisions.

In hierarchical spectrum sensing, instead of being in the silent mode when $\lambda_1 < \Lambda_i < \lambda_2$, cooperative nodes forward their observed energy values (e.g., in bit form) to the fusion center as shown in Fig. 3.1c [10, 27]. The fusion center applies soft combing for the energy values and the hard combining for the decision values D_i's. The double threshold strategy can increase the available degrees of freedom.

The double threshold energy detector performs better than the conventional energy detector in cooperative spectrum sensing networks with an OR rule. On the other hand, the conventional energy detector performs better than the double threshold energy detector in multi-user networks operated non-cooperatively [19, 23, 27, 33].

3.3 Blindly Combined Energy Detector

While the conventional energy detector is optimal for independent signal samples, received signal samples may be correlated because transmitted signals, channels and/or noises may have physical connections. In this case, the blindly combined energy detection (BCED), which outperforms conventional energy detector for correlated signals, is proposed in [30].

Since BCED processes correlated signals, a receiver with M antennas ($M \geq 1$) is considered. Thus, the nth received signal sample at the ith antenna can be given as $y_i(n) = w_i(n)$ and $y_i(n) = x_i(n) + w_i(n)$ under hypotheses $\mathcal{H}_0$ and $\mathcal{H}_1$, respectively. Then, the received signal from M antennas in N time slots can be stacked to a vector $\mathbf{r}(n)$ with size $NM \times 1$ as

$$\mathbf{r}(n) = [\mathbf{y}_1(n-1) \cdots \mathbf{y}_M(n-1) \cdots \mathbf{y}_1(n-N) \cdots \mathbf{y}_M(n-N)]^T$$

$$= \begin{cases} \mathbf{w}(n) & : \mathcal{H}_0 \\ \mathbf{x}(n) + \mathbf{w}(n) & : \mathcal{H}_1 \end{cases} \tag{3.6}$$

where $\mathbf{x}(n)$ and $\mathbf{w}(n)$ are also vectors with size $NM \times 1$, $n = 1, \cdots, N$, and superscript T means transpose operation. Covariance matrices of $\mathbf{x}(n)$ and $\mathbf{r}(n)$ are given as $\mathbf{R_x} = \mathbb{E}[\mathbf{x}(n)\mathbf{x}^T(n)]$ and $\mathbf{R_r} = \mathbb{E}[\mathbf{r}(n)\mathbf{r}^T(n)]$, respectively. If eigenvector $\boldsymbol{\beta}$ with size $NM \times 1$ is corresponding to the maximum eigenvalue of $\mathbf{R_x}$, the optimal combing that can maximize the receive SNR is $\mathbf{z}(n) = \boldsymbol{\beta}^T \mathbf{r}(n)$, i.e., the received signal vector combined with the eigenvector $\boldsymbol{\beta}$. Then the test statistics of optimally combined energy detection can be given as

$$\Lambda_{BCED} = \sum_{n=1}^{N} |\mathbf{z}(n)|^2 . \tag{3.7}$$

In energy detection, since the receiver does not have prior knowledge of transmit signal, $\mathbf{R_x}$ is unknown to the receiver. Since $\mathbf{R_r}$ and $\mathbf{R_x}$ have the same eigenvectors, $\boldsymbol{\beta}$ can be found approximately by the eigenvector corresponding to the maximum eigenvalue of the sample covariance matrix of received signal vector, i.e., $\tilde{\mathbf{R}}_r(N) = \frac{1}{N} \sum_{n=0}^{N-1} \mathbf{r}(n)\mathbf{r}^T(n)$. In this case, signals can be combined without any a priori information of the source signal and channel, which is referred to as *blindly combining*.

In [30], it further shows that the BCED does not need the noise variance estimation. Therefore, performance degradation due to noise uncertainty can be overcome. Detection capability of BCED is much better than the conventional energy detector for highly correlated signals. In addition, the received SNR of BCED signal is higher than that of traditional energy detector because of the operation $\mathbf{z}(n) = \boldsymbol{\beta}^T \mathbf{r}(n)$.

3.4 Energy Detector with an Arbitrary Power Operation

The improved energy detector and the L_p-Norm energy detector use arbitrary power operation in energy calculations.

3.4.1 Improved Energy Detector

The conventional energy detector maximizes the generalized likelihood function (2.4), which may not maximize the detection probability or minimize the false-alarm/missed detection probability. This motivates an improved energy detector, where test statistic is formed by replacing the squaring operation of the conventional energy detector with an arbitrary positive power [2]. The test statistic of improved energy detector is thus given as

$$\Lambda = \sum_{n=1}^{N} |\mathbf{y}(n)|^p \tag{3.8}$$

where p is an arbitrary positive value. When $p = 2$, this detector is equivalent to the conventional energy detector.

For a Gaussian random signal (**S2** signal model) corrupted with the Gaussian noise, unlike the $p = 2$ case, the distribution of the test statistics (3.8) is not a gamma (or a central chi-square) distribution in general. The derivation of the exact distribution in general case is complicated. Therefore, a gamma approximation is developed by matching mean, $\mathbb{E}[\Lambda|\mathscr{H}_i]$, and variance, $\mathbb{V}\mathrm{ar}[\Lambda|\mathscr{H}_i]$ of the test statistic to scaling factor, k_i, and shaping factor, θ_i, of the gamma distribution as

$$k_i = \frac{\mathbb{E}[\Lambda|\mathscr{H}_i]^2}{\mathbb{V}\mathrm{ar}[\Lambda|\mathscr{H}_i]}; \quad \theta_i = \frac{\mathbb{V}\mathrm{ar}[\Lambda|\mathscr{H}_i]}{\mathbb{E}[\Lambda|\mathscr{H}_i]} \tag{3.9}$$

where $i \in \{0,1\}$. The approximation is better for small p, high N and low average SNR. The accuracy in low SNR is important for practical cognitive radio networks because their signal levels are usually very weak.

The optimum power value depends on the target probabilities (P_f and P_d), average SNR $\bar{\gamma}$, and number of samples N. In a practical system, if P_f, P_d, and N are pre-defined, and $\bar{\gamma}$ is estimated by an SNR estimation method, the optimum p can be calculated from the receiver operating characteristic (ROC) curve which is given as

$$P_d = 1 - F_{\Lambda|\mathscr{H}_1}\left(F_{\Lambda|\mathscr{H}_0}^{-1}(P_f, k_0, \theta_0), k_1, \theta_1\right) \tag{3.10}$$

where $F_{\Lambda|\mathscr{H}_i}(\cdot)$ and $F_{\Lambda|\mathscr{H}_i}^{-1}(\cdot)$ are CDF and inverse CDF, respectively, of gamma distribution under hypothesis $\mathscr{H}_i$.

The analysis in [2] shows that the improved energy detector outperforms the conventional energy detector, especially in low SNR. However both detectors have similar performance at high SNR. Further analysis in [2] shows that the optimum p approaches a common floor value at high SNR when P_d is maximized at a fixed P_f. Conversely, the optimum p approaches a common floor value at low SNR when P_f is minimized at a fixed P_d. Moreover, an improved energy detector is considered for a cooperative spectrum sensing network (multiple antennas) in [22], where the test statistic of the ith antenna is $\sum_{n=1}^{N} |\mathbf{y}_i(n)|^p$.

3.4.2 L_p-Norm Detector

While the conventional energy detector is optimal for Gaussian noise, for non-Gaussian noise, it incurs a significant performance loss. Non-Gaussian noise and interference are typically modeled as Gaussian mixture noise (GMN) in which ϵ-mixture noise is a special case, the generalized Gaussian noise (GGN) in which Laplacian and AWGN are special cases, and co-channel interference (CCI).

Binary hypothesis problem given in (2.1) is valid with any noise model. Although the maximum likelihood test is globally optimal for the hypothesis test, it is quite cumbersome to evaluate with different noise models. Moreover, it requires knowledge of the distribution of the unknown signal and also knowledge of the noise distribution, which may be difficult to estimate because they change with time in practice.

Motivated by non-Gaussian noise environment and other practical difficulties, a suboptimal L_p-Norm detector is proposed in [13, 14]. In general, signal model **S3** is considered. At low SNR, a locally optimal test statistic is approximated as

$$\Lambda_{LO} \cong \frac{\sigma_h^2}{4} \sum_{n=1}^{N} \mathbf{g}(\mathbf{y}(n)) \tag{3.11}$$

where $\sigma_h^2 = \mathbb{E}[|\mathbf{h}|^2]$, $\mathbf{g}(\mathbf{y}(n)) = \dfrac{\left(\frac{\partial^2}{\partial y_r^2} + \frac{\partial^2}{\partial y_i^2}\right) f_\mathbf{w}(\mathbf{y}(n))}{f_\mathbf{w}(\mathbf{y}(n))}$, $f_\mathbf{w}(w)$ is the noise PDF, and $\mathbf{y}_r$ and $\mathbf{y}_i$ are real and imaginary components of $\mathbf{y}$. If $\mathbf{w}$ is a Gaussian noise, Λ_{LO} is equivalent to the test statistic of the conventional energy detector given in (2.5). Although (3.11) does not require the distribution of the transmitted signal, it needs the noise distribution to evaluate non-linear function $\mathbf{g}(\mathbf{y}(n))$. Thus, by replacing $\mathbf{g}(\mathbf{y}(n))$ with a simpler function which is not a function of noise distribution, the L_p-Norm detector is proposed [13,14]. The test statistic of the suboptimal L_p-Norm detector is given as[1]

$$\Lambda_{SO} = \sigma_h^2 \sum_{n=1}^{N} |\mathbf{y}(n)|^p \tag{3.12}$$

where p is a tunable parameter with a positive value.

The L_2-Norm detector is the conventional energy detector. The L_p-Norm detector is a more general detector than the improved energy detector which has been analyzed only for the Gaussian noise. By optimizing the value p, L_p-Norm detector may achieve better detection performance for a large class of different noise and interference models.

The distributions of Λ_{SO} under hypotheses $\mathcal{H}_0$ and $\mathcal{H}_1$, denoted as $f_{\Lambda_{SO}|\mathcal{H}_0}(x)$ and $f_{\Lambda_{SO}|\mathcal{H}_1}(x)$, respectively, can also be approximated as Gaussian distributions (CLT approach). The mean and variance under $\mathcal{H}_0$ can be easily calculated, and thus P_f calculation is straightforward. Since the mean and variance under $\mathcal{H}_1$ depend on the channel, signal, and noise, derivation of an accurate expression of P_d by using CLT is still difficult. Therefore, the average detection probability may be evaluated numerically for different noise and interference models.

For example, the variation of missed-detection probability P_{md} versus p is shown in Fig. 3.2 for different noise and interference models. The false-alarm probability is set to be 0.1. The lowest missed-detection probability for AWGN, heavy-tailed ϵ-mixture noise or short-tailed GGN is at $p = 2$, $p < 2$ or $p > 2$, respectively [13]. This means that the L_2-Norm detector is not optimal for a non-Gaussian noise.

---

[1]The number of diversity branches L which is considered in [13, 14] is neglected here.

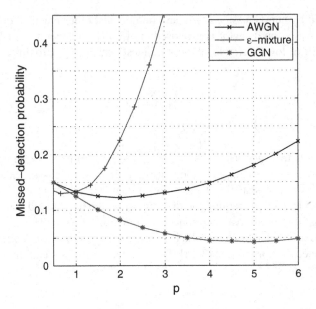

Fig. 3.2 The missed-detection probability versus p for the L_p-Norm detector with $\gamma = -13$ dB and $P_f = 0.1$

3.5 Hybrid/Coherent Energy Detection

If primary user uses pilot symbols for the synchronization and channel estimation purposes, denoted by $\mathscr{P}$ and $\mathscr{D}$ as the set of positions of pilot symbols and data symbols, respectively, these pilots can be exploited in coherent spectrum sensing. This technique significantly improves the detection performance by removing noise/interference uncertainties at low SNR.

Given $\mathscr{P}$ and $\mathscr{D}$, an optimal hybrid detector can be derived using likelihood ratio test [24]. However, this detection metric requires high computational complexity. In practice, the receiver may not know the exact knowledge of the data symbols (e.g., due to adaptive modulation).

Due to high complexity of this detector, a locally optimal hybrid/coherent energy detector which is optimal in low SNR and sub-optimal in moderate and high SNRs is introduced in [15, 16]. This hybrid detection metric is given as[2]

$$\Lambda_{LO} = \frac{\sigma_h^2}{\sigma_w^2} \sum_{n \in \mathscr{P}} |\mathbf{y}(n)\mathbf{s}^*(n)|^2 + \frac{\sigma_s^2 \sigma_h^2}{\sigma_w^2} \sum_{n \in \mathscr{D}} |\mathbf{y}(n)|^2 \qquad (3.13)$$

---

[2]The number of diversity branches L which is considered in [16] is neglected here.

where $\mathbf{s}^*(n)$ is the complex conjugate of $\mathbf{s}(n)$. The first term is a simple linear combination of a coherent metric which involves the correlation of the symbols received at the pilot positions and the pilot which is known to the receiver. The second term is a simple non-coherent energy detection metric which includes unknown data symbols' positions. As shown in [15], the hybrid detector outperforms the conventional energy detection even pilot data positions are not known a priori.

3.6 Sequential Energy Detection

The goal of sequential based approaches is to reduce the sensing time for a given target performance [9, 28]. Unlike the conventional energy detector which always waits for a pre-decided number of samples to be received before calculating the test statistic, in the sequential energy detection, the samples are accepted sequentially. At each time step i, the likelihood ratio is calculated [8]. The sequential based approaches use sequential probability ratio test (SPRT) to calculate the likelihood ratio at each time step until either lower or upper threshold is satisfied. These lower and upper thresholds are computed using the Wald-Wolfowitz theorem [26]. The log-likelihood ratio can be updated as

$$\ln \Delta_{i+1} = \ln \Delta_i + \ln \zeta + \eta \mathbf{y}(i+1)\mathbf{y}^*(i+1)$$

where $\Delta_i = \zeta^i e^{\eta \sum_{j=1}^i \mathbf{y}(j)\mathbf{y}^*(j)}$, $\zeta = \frac{\sigma_w^2}{\sigma_w^2 + \sigma_s^2}$ and $\eta = \frac{\sigma_s^2}{2\sigma_w^2(\sigma_w^2 + \sigma_s^2)}$.

This model avoids the approximations made by some other sequential based approaches (e.g., [28]). Since sequential energy detector does not wait for the whole sensing time, it can achieve a significant throughput improvement over the conventional energy detector (approximately 2–6 times), and also achieve a good performance improvement in low SNR region.

3.6.1 Doubly Sequential Energy Detection

The doubly sequential energy detector performs sequential detection at individual cooperative nodes and the fusion center. In cooperative sensing, the sequential test is performed at the fusion center based on cooperative nodes' soft messages, or all nodes in the network optimize a joint detection as distributed manner. This minimizes the sensing overhead, which is the time required for both sequential testings at the cooperative nodes and the fusion center, while achieving the specified detection performance.

Since nodes can perform sequential test independently, the termination criterion at the fusion center is vital. Considering the time taken for the decision at the fusion center, several decision criteria are introduced in [7]:

- The fusion center simply takes the decision of the first reporting node, which is called *one shot* detection. This greatly reduces the sensing time but it has poor performance in terms of P_f.
- The fusion center continues the test till it has received M decisions in favor of a particular hypothesis, which is called *first-M-positive* detection.
- After the fusion center receives the first decision, it waits another time period, T_{th}, and the final decision is made according to the majority rule among the all received decisions which is called *wait-till-T_{th}* detection.

The first-M-positive detection outperforms others in terms of the tradeoff between sensing time and detection performance.

3.7 Adaptive Detection

Adaptive algorithms are employed when the conventional energy detector may fail due to insufficient signal strength. This happens when the signal bandwidth is narrow compared to the detector window, or when only a small fraction of the signal bandwidth is captured within the sensing window. In either case, signal energy does not meet pre-defined threshold level. Some possible ways to combat this problem are:

- lower the threshold, which increases P_f significantly;
- choose a window matching to the signal, which needs a priori knowledge of primary signal;
- narrow the detector window, which increases the sensing time.

Due to the drawbacks of performance and implementation complexity of the first two solutions, the third solution is developed as an adaptive energy detector [29]. The window size is selected adaptively by subdividing the original window size.

In dynamic wireless communication networks, the primary signal can appear or disappear at any time instant within the sensing time. These intermediate state changes are taken into account in the adaptive structure proposed in [21], in which a side detector is implemented to monitor continuously the presence of the primary signal within sensing time. The side detector, which is also an energy detector, produces a fast and rough binary decision about primary signal state. When side detector identifies primary signal presence, it sends a trigger to the adaptive energy detector to set the threshold adaptively. However, the implementation of side detector is not explained, and thus advantage of performance improvement is not clear, as well as the implementation overhead.

A different adaptive structure is proposed in [3]. In contrast to the periodic sensing in conventional energy detector, this adaptive sensing technique performs channel sensing only when it is needed, and therefore, unnecessary sensing can be avoided. It makes each decision based on the previous sensing results, and thus, the sensing time is much shorter than that of the conventional energy detector. A shorter sensing period enhances the adaptability by increasing the frequency of decision intervals. For more reliability, multiple soft sensing results can be combined to generate reliable detection.

3.8 Generalized Energy Detector

The conventional energy detector is not optimal when noise samples are non-Gaussian and/or not independent. Thus, generalized energy detectors are introduced focusing on two cases:

- when noise samples are non-Gaussian and independent
- when noise samples are Gaussian but not independent.

The pre-whitening transformation together with non-linear functions can be used to produce independent and Gaussian samples [18, 20, 25].

For non-Gaussian and independent noise case, a non-linear function can be used to convert the noise variable w having arbitrary distribution function $F_w(w)$ to a zero-mean and unit-variance Gaussian random variable. For example, such a function is defined in [17] as

$$g(w) = \phi^{-1}\left(F_w(w)\right) \text{ where } \phi(x) = \int_{-\infty}^{x} \frac{e^{-\frac{t^2}{2}}}{\sqrt{2\pi}} dt. \qquad (3.14)$$

If the above transformation is applied to every received signal sample of $\mathbf{y}$, the noise terms in the resultant set of samples become independent Gaussian random variables. Subsequently, the energy can be calculated by using the test statistic of conventional energy detection (2.5). This energy detector is called an extended energy detector.

For Gaussian but not independent noise case, the independence is obtained by using a matrix transformation (linear) derived from independent component analysis (ICA). First, an uncorrelated vector of received signal samples $\mathbf{y}$ is generated, and then independent samples are obtained by using an unitary transformation. These two steps can be combined into a single transformation $\mathbf{U}$. For example, such a transformation is defined in [17] as

$$\mathbf{U} = \mathbf{Q}\mathbf{R}_w^{-1/2} \qquad (3.15)$$

where $\mathbf{Q}^T\mathbf{Q} = \mathbf{I}$ ($\mathbf{I}$ is unitary), and $\mathbf{R}_w = \mathbb{E}[\mathbf{w}\mathbf{w}^T]$ which is the noise covariance matrix.

By using both non-linear function and transformations given in (3.14) and (3.15), the test statistic of an energy detector which may receive non-Gaussian and non-independent noise is given as

$$\Lambda = g\left(\mathbf{Q}\mathbf{R}_w^{-1/2}\mathbf{y}\right)^T g\left(\mathbf{Q}\mathbf{R}_w^{-1/2}\mathbf{y}\right)$$

which has a monotonic value with the integrator output of the conventional energy detector. Thus, this detector is also called a pre-processed extended energy detector [17]. Further, $\mathbf{U}$ and $F_w(w)$ can be estimated by using training and a non-parametric model, which may increase implementation complexity. In addition, this detector does not guaranty optimality.

3.9 Other Energy Detectors

Apart from the aforementioned energy detectors, there are other alternative methods.

- Sliding window energy detector: Power is accumulated in window length L which may slide over a specified frequency range [6]. The power versus frequency plot generates a smooth curve from which the noise can be identified from the signal, and transmitted signal in different frequency bands can also be distinguished. While the detection probability increases with the increment of window length L, spectral efficiency decreases.
- Censored energy detector: While cooperative spectrum sensing achieves quick and reliable detection [5, 31], cooperative users report their sensing results to a fusion center, with either soft or hard decision manner. However, to make a final decision reliably, all decisions from all cooperative nodes may not be required because some cooperative nodes' decisions may be incorrect due to severe fading and background noise. Therefore, a censored energy detector [1] compares the measurement of each cooperative node with two pre-determined limits, and only measurements that are smaller than the lower limit (Λ_1) or larger than the upper limit (Λ_2) are reported to the fusion center. This detection strategy saves transmission power and network overhead. The performance of this method depends on the thresholds, decision rule (soft or hard decision rule), operating SNR and the number of collaborating users.

References

1. Chen, Y. (2010) , Analytical performance of collaborative spectrum sensing using censored energy detection. IEEE T on Wireless Communications 9(12): 3856–3865.
2. Chen, Y. (2010) Improved energy detector for random signals in Gaussian noise. IEEE T on Wireless Communications 9(2): 558–563.

3. Choi, K. W. (2010) Adaptive sensing technique to maximize spectrum utilization in cognitive radio. IEEE T on Vehicular Technology **59**(2): 992–998.
4. Ding, J., Guo, J., Qu, D., Jiang, T. (2009) Novel multiple slots energy detection for spectrum sensing in cognitive radio networks. In: Proceedings of Asia-Pacific Conference on Communications (APCC), Shanghai, 8–10 Oct 2009.
5. Ganesan, G., Li, Y. (2007) Cooperative spectrum sensing in cognitive radio, part I: Two user networks. IEEE T on Wireless Communications **6**(6): 2204–2213.
6. Kim, Y. M., Zheng, G., Sohn, S. H., Kim, J. M. (2008) An alternative energy detection using sliding window for cognitive radio system. In: Proceedings of International Conference on Advanced Communications Technology (ICACT), Pyeongchang, 17–20 Feb 2008.
7. Kundargi, N., Tewfik, A. (2010) Doubly sequential energy detection for distributed dynamic spectrum access. In: Proceedings of IEEE International Conference on Communications (ICC), Cape Town, 23–27 May 2010.
8. Kundargi, N., Tewfik, A. (2010) A performance study of novel sequential energy detection methods for spectrum sensing. In: Proceedings of IEEE International Conference on Acoustics, Speech, and Signal Processing (ICASSP), Texas, 14–19 March 2010.
9. Lai, L., Fan, Y., Poor, H. V. (2008) Quickest detection in cognitive radio: A sequential change detection framework. In: Proceedings of IEEE Global Telecommunications Conference (GLOBECOM), New Orleans, 30 Nov-4 Dec 2008.
10. Liu, S. Q., Hu, B. J., Wang, X. Y. (2012) Hierarchical cooperative spectrum sensing based on double thresholds energy detection. IEEE Communications Letters **16**(7): 1096–1099.
11. Ma, J., Li, Y. (2008) A probability-based spectrum sensing scheme for cognitive radio. In: Proceedings of IEEE International Conference on Communications (ICC), Beijing, 19–23 May 2008.
12. Ma, J., Zhou, X., Li, G. (2010) Probability-based periodic spectrum sensing during secondary communication. IEEE T on Communications **58**(4): 1291–1301.
13. Moghimi, F., Nasri, A., Schober, R. (2009) L_p-norm spectrum sensing for cognitive radio networks impaired by Non-Gaussian noise. In: Proceedings of IEEE Global Telecommunications Conference (GLOBECOM), Hawaii, 30 Nov- 4 Dec 2009.
14. Moghimi, F., Nasri, A., Schober, R. (2011) Adaptive L_p-norm spectrum sensing for cognitive radio networks. IEEE T on Communications **59**(7): 1934–1945.
15. Moghimi, F., Schober, R., Mallik, R. K. (2010) Hybrid coherent/energy detection for cognitive radio networks, In: Proceedings of IEEE Global Telecommunications Conference (GLOBECOM), Miami, 6–10 Dec 2010.
16. Moghimi, F., Schober, R., Mallik, R. K. (2011) Hybrid coherent/energy detection for cognitive radio networks. IEEE T on Wireless Communications **10**(5): 1594–1605.
17. Moragues, J., Vergara, L., Gosálbez, J., Bosch, I. (2009) An extended energy detector for non-Gaussian and non-independent noise. Signal Processing **89**(4): 656–661.
18. Nicolas, P., Kraus, D. (1988) Detection and estimation of transient signals in coloured Gaussian noise. In: Proceedings of International Conference on Acoustics, Speech, and Signal Processing (ICASSP), New York, 11–14 Apr 1988.
19. Sardellitti, S., Barbarossa, S., Pezzolo, L. (2009) Distributed double threshold spatial detection algorithms in wireless sensor networks. In: IEEE 10th Workshop on Signal Processing Advances in Wireless Communications (SPAWC), Perugia, 21–24 June 2009.
20. Schultheiss, P. M., Godara, L. C. (1994) Detection of weak stochastic signals in non-gaussian noise: a general result. In: Proceedings of International Conference on Acoustics, Speech, and Signal Processing (ICASSP), Adelaide, 19–22 Apr 1994.
21. Shehata, T. S., El-Tanany, M. (2009) A novel adaptive structure of the energy detector applied to cognitive radio networks. In: Proceedings of 11th Canadian Workshop on Information Theory (CWIT), Ottawa, 13–15 May 2009.

22. Singh, A., Bhatnagar, M. R., Mallik, R. K. (2012) Cooperative spectrum sensing in multiple antenna based cognitive radio network using an improved energy detector. IEEE Communications Letters **16**(1): 64–67.
23. Sun, C., Zhang, W., Letaief, K. B. (2007) Cooperative spectrum sensing for cognitive radios under bandwidth constraints. In: IEEE Wireless Communications and Networking Conference (WCNC), Hong Kong, 11–15 Mar 2007.
24. Trees, H. L. V. (2001) Detection, Estimation, and Modulation Theory, Part I, Wiley-Interscience.
25. Urkowitz, H. (1969) Energy detection of a random process in colored Gaussian noise. IEEE T on Aerospace and Electronic Systems **AES-5**(2): 156–162.
26. Wald, A. (2004) Sequential Analysis (Dover Phoenix Editions), Dover Publications.
27. Wu, J., Luo, T., Li, J., Yue, G. (2009) A cooperative double-threshold energy detection algorithm in cognitive radio systems. In: Proceedings of 5th International Conference on Wireless Communications, Networking and Mobile Computing (WiCOM), Beijing, 24–26 Sept 2009.
28. Xin, Y., Zhang, H., Rangarajan, S. (2009) SSCT: A simple sequential spectrum sensing scheme for cognitive radio. In: Proceedings of IEEE Global Telecommunications Conference (GLOBECOM), Hawaii, 30 Nov- 4 Dec 2009.
29. Xu, J. Y., Alam, F. (2009) Adaptive energy detection for cognitive radio: An experimental study. In: 12th International Conference on Computers and Information Technology (ICCIT), Dhaka, 21–23 Dec 2009.
30. Zeng, Y., Liang, Y. C., Zhang, R. (2008) Blindly combined energy detection for spectrum sensing in cognitive radio. IEEE Signal Processing Letters **15**: 649–652.
31. Zhang, W., Mallik, R., Letaief, K. B. (2009) Optimization of cooperative spectrum sensing with energy detection in cognitive radio networks. IEEE T on Wireless Communications **8**(12): 5761–5766.
32. Zhou, X., Ma, J., Li, G. Y., Kwon, Y. H., Soong, A. C. K. (2009) Probability-based optimization of inter-sensing duration and power control in cognitive radio. IEEE T on Wireless Communications **8**(10): 4922–4927.
33. Zhu, J., Xu, Z., Wang, F., Huang, B., Zhang, B. (2008) Double threshold energy detection of cooperative spectrum sensing in cognitive radio. In: Proceedings of International Conference on Cognitive Radio Oriented Wireless Networks and Communications (CrownCom), Singapore, 15–17 May 2008.

Chapter 4
Performance Measurements

The energy detection process is affected by the fluctuations of the propagation channel, which occur due to path loss, large-scale fading and small-scale fading. Channel correlation is also a major impact when energy detectors are implemented in cooperative spectrum sensing networks [43].

Before 1980s, energy detection and its performance were studied under different network conditions [16,18,25,26,31,48,58,59]. As energy detection is being applied to new wireless technologies, its applications and performance have been revisited recently for many fading channels and diversity reception schemes [4, 7, 15, 23, 24, 32–34, 37, 47].

The performance of the energy detector may be characterized by using some of the following metrics:

- false-alarm probability
- detection or missed-detection probability
- receiver operating characteristic (ROC) curve
- sensing gain
- total error rate
- area under ROC curve (AUC).

The false-alarm, detection and missed-detection probabilities are the basic performance metrics. In the conventional direct communications (non-cooperative networks), the false alarm probability, which is for the case of absence of the signal, depends only on number of samples, threshold, and noise power but not on the wireless channel. Therefore, it can be easily calculated by using (2.11) because there is no fading/shadowing effect incurred by the wireless channel.

In the following, other performance metrics are discussed for **S1** signal model, unless otherwise specified.

S. Atapattu et al., *Energy Detection for Spectrum Sensing in Cognitive Radio*,
SpringerBriefs in Computer Science, DOI 10.1007/978-1-4939-0494-5_4,
© The Author(s) 2014

4.1 Average Detection Probability

The detection probability, P_d, which is for the case of presence of signal, depends not only on number of samples, threshold, and noise power but also on the wireless channel. By using (2.13), an instantaneous $P_d(\gamma)$ (the detection probability corresponding to SNR γ) can be evaluated. But it is essential to evaluate the average detection probability, $\overline{P_d}$, due to fading/shadowing effect of the wireless channel. This cannot be calculated straightforwardly because the instantaneous $P_d(\gamma)$ should be averaged over the channel SNR distribution, $f_\gamma(\gamma)$. Some possible approaches of finding $\overline{P_d}$ are discussed in the following.

4.1.1 Direct Averaging

The average detection probability, $\overline{P_d}$, can be calculated by direct averaging of $P_d(\gamma)$ over the respective SNR distribution as

$$\overline{P_d} = \int_0^\infty P_d(x) f_\gamma(x)\,dx. \tag{4.1}$$

This is the traditional way of deriving $\overline{P_d}$ [23,24,35,37]. The derivation requires the generalized Marcum-Q function to be averaged over the SNR distribution. As only limited integral results are available, mathematical complexity is inevitable with Marcum-Q function. A set of closed-form integrals, approximations, and bounds for the generalized Marcum-Q function is available in [3,39,41,44,45,52–54].

For a Rayleigh fading channel, i.e., $f_\gamma(x) = \frac{1}{\bar{\gamma}} e^{-x/\bar{\gamma}}$ where $\bar{\gamma}$ is the average SNR, $\overline{P_d}$ is derived as [24]

$$\overline{P_d} = e^{-\frac{\lambda}{2\sigma_w^2}} \sum_{p=0}^{N-2} \frac{1}{p!} \left(\frac{\lambda}{2\sigma_w^2}\right)^p$$

$$+ \left(\frac{1+N\bar{\gamma}}{N\bar{\gamma}}\right)^{N-1} \left[e^{-\frac{\lambda}{2\sigma_w^2(1+N\bar{\gamma})}} - e^{-\frac{\lambda}{2\sigma_w^2}} \sum_{p=0}^{N-2} \frac{1}{p!} \left(\frac{\lambda N\bar{\gamma}}{2\sigma_w^2(1+N\bar{\gamma})}\right)^p \right] \tag{4.2}$$

where N is the number of samples or the time-bandwidth product. Closed-form expressions of $\overline{P_d}$ over Rician, Nakagami-m, generalized K, and mixture gamma channel models are also given in [7,23,24,35,37].

4.1.2 MGF Approach

Since the generalized Marcum-Q function can be written as a circular contour integral within the contour radius $r \in [0, 1)$, (2.13) can be re-written as [57]

$$P_d(\gamma) = \frac{e^{-\frac{\lambda}{2\sigma_w^2}}}{j2\pi} \oint_\Omega \frac{e^{(\frac{1}{z}-1)N\gamma + \frac{\lambda}{2\sigma_w^2}z}}{z^N(1-z)} dz \qquad (4.3)$$

where Ω is a circular contour of radius $r \in [0, 1)$. Since the moment generating function (MGF) of received SNR γ is $\mathcal{M}_\gamma(s) = \mathbb{E}(e^{-s\gamma})$, the average detection probability, $\overline{P_d}$, can be given as [10, 35]

$$\overline{P_d} = \frac{e^{-\frac{\lambda}{2\sigma_w^2}}}{j2\pi} \oint_\Omega \mathcal{M}_\gamma\left[\left(1 - \frac{1}{z}\right)N\right] \frac{e^{\frac{\lambda}{2\sigma_w^2}z}}{z^N(1-z)} dz. \qquad (4.4)$$

Since the Residue Theorem [38] in complex analysis is a powerful tool to evaluate line integrals and/or real integrals of functions over closed curves, (4.4) may be solved in closed-form for mathematically tractable MGFs. Since $\mathcal{M}_\gamma(s) = \frac{1}{1+\bar{\gamma}s}$ for Rayleigh fading channels, $\overline{P_d}$ can be derived as

$$\overline{P_d} = \begin{cases} e^{-\frac{\lambda}{2\sigma_w^2}}\left(\text{Res}\,(g;0) + \text{Res}\left(g; \frac{N\bar{\gamma}}{1+N\bar{\gamma}}\right)\right) : N > 1 \\ e^{-\frac{\lambda}{2\sigma_w^2(1+\bar{\gamma})}} : \qquad\qquad\qquad\qquad N = 1 \end{cases} \qquad (4.5)$$

where $\text{Res}\,(g;0)$ and $\text{Res}\big(g; N\bar{\gamma}/(1+N\bar{\gamma})\big)$ denote the residues of the function $g(z)$ at the origin and at $z = N\bar{\gamma}/(1 + N\bar{\gamma})$, respectively, and

$$g(z) = \frac{e^{\frac{\lambda}{2\sigma_w^2}z}}{(1 + N\bar{\gamma})z^{(N-1)}(1 - z)(z - \frac{N\bar{\gamma}}{1+N\bar{\gamma}})}.$$

By using MGF approach, closed-form expressions for $\overline{P_d}$ are derived over Rician, Nakagami-m, η-μ channel models, and relay channels in [4, 5, 34]. If $\mathcal{M}_\gamma(s)$ is in a simple rational form (e.g., Nakagami-m and η-μ fading), MGF approach based on residue evaluation is effective. Unfortunately, MGFs of some fading models (e.g., K or K_G model [7]) or some network scenarios (e.g., cooperative spectrum sensing) do not give rise to a rational-form MGF. For those scenarios, the following suggestions can be applicable:

- In [12], a mixture gamma (MG) model is proposed for the distribution of the SNR, which can accurately approximate most existing fading channels. The MGF of the MG model is in a simple rational form.

- In [10], the rational form MGFs are derived for cooperative spectrum sensing in cognitive radio.
- In general, the Taylor series and the Padé approximation of the MGF can generate rational forms.

If the exact MGF is not in a suitable rational form, an accurate rational approximation may be derived by using one of the above methods, which can be used in (4.4). Thus, this analytical approach based on the MGF furnishes a unified framework.

4.1.3 Infinite Series Representation

Deriving a closed-form expression for $\overline{P_d}$ is challenging when slow fading is considered, e.g., log-normal shadowing. Due to lack of direct integral results of the Marcum-Q function, alternative approaches are proposed in [2,35,56] using infinite series expansions.

By using [54, Eq. (4.63)], P_d can be given as [35]

$$P_d(\gamma) = 1 - e^{-\frac{\lambda}{2\sigma_w^2}} \sum_{n=N}^{\infty} \left(\frac{\lambda}{2\sigma_w^2 N\gamma} \right)^n e^{-N\gamma} I_n \left(\sqrt{\frac{2\lambda N\gamma}{\sigma_w^2}} \right) \qquad (4.6)$$

where $I_n(\cdot)$ is the nth order modified Bessel function of the first kind. Although $\overline{P_d}$ can be derived with a Hypergeometric function of two variables, a series truncation is required for the numerical calculation of this infinite summation.

By using [28, Eq. (8.445)] and [54, Eq. (4.60)], $P_d(\gamma)$ can also be given in an infinite series expression as [2,56]

$$P_d(\gamma) = \sum_{n=0}^{\infty} \frac{N^n \Gamma \left(N+n, \frac{\lambda}{2\sigma_w^2} \right) \gamma^n e^{-N\gamma}}{n! \Gamma(N+n)}. \qquad (4.7)$$

Since $\int_0^\infty x^n e^{-Nx} f_\gamma(x) = (-1)^n \mathscr{M}_\gamma^{(n)}(s)\big|_{s=N}$ where $\mathscr{M}_\gamma^{(n)}(s)$ is the nth derivative of MGF of γ with respect to s, $\overline{P_d}$ can be evaluated over a generalized fading channel as

$$\overline{P_d} = \sum_{n=0}^{\infty} \frac{(-1)^n N^n \Gamma \left(N+n, \frac{\lambda}{2\sigma_w^2} \right)}{n! \Gamma(N+n)} \mathscr{M}_\gamma^{(n)}(s)\big|_{s=N}. \qquad (4.8)$$

This representation facilitates derivation of $\overline{P_d}$ with a non-integer fading parameter (e.g., m in Nakagami-m fading channel). In [56], an approximation of $\overline{P_d}$ is derived over a slow fading channel by replacing the log-normal distribution with a Wald distribution or inverse Gaussian distribution [36]. However, this approximation becomes loose at the right tail of the log-normal distribution specially when the shadowing standard deviation is high. Moreover, $\overline{P_d}$ can be derived by using MGF of γ [2] or Laguerre polynomial [46] which also has the capability of providing results for fractional fading parameter m. When infinite series representations are used, series truncation may be required for the numerical calculation. Different infinite series representations have different convergence rates.

The average detection probability versus the average SNR (for a given false alarm probability or a given threshold) is one way of representing the detection performance of an energy detector. In Fig. 4.1, the average missed-detection probability ($\overline{P_{md}} = 1 - \overline{P_d}$) versus the average SNR is shown for Nakagami-m fading channel where $m = 1, 2, 3$. The energy detector has better performance for higher N, m or $\bar{\gamma}$. Moreover, the value m shows how fast the average missed-detection probability decreases with the increase in the SNR in the high SNR region. This observation motivates the sensing gain to be discussed in Sect. 4.2.

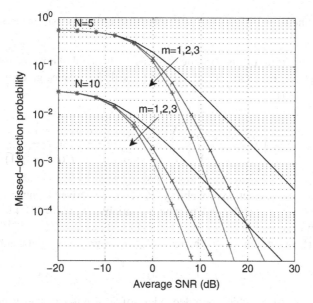

Fig. 4.1 The average missed-detection probability with average SNR of **S1** signal model for $m = 1$ (Rayleigh), 2, and 3 in Nakagami-m fading channel with $2\sigma_w^2 = 1$, $\lambda = 5$, and $N = 5$ or 10

4.2 Sensing Gain

Although closed-form solutions for $\overline{P_d}$ are useful, their derivation may not be tractable for every scenario. Moreover, rigorously derived formulas include special functions, which are not explicit about impact of fading/shadowing/diversity. In this case, series expansion (usually infinite series expansion) or an asymptotic analysis (usually at high SNR approximations) can be used to obtain approximate and asymptotic expressions, which can provide more meaningful results/insights than rigorous formulas. This motivates to introduce the sensing gain by using an asymptotic analysis [60]. The sensing gain quantifies the impact of channel fading and spatial diversity.

A general series expansion of the missed-detection probability is given under two reasonable assumptions:

- the instantaneous SNR $\gamma = \beta\bar{\gamma}$ where β is a channel dependent non-negative random variable.
- the PDF of β well behaves at $\beta = 0$ as

$$f_\beta(\beta) = \sum_{j=0}^{\infty} a_j \beta^{t+j} \tag{4.9}$$

where a_j and t can be evaluated using the Maclaurin expansion of the exact PDF of $f_\gamma(\gamma)$.

This is very similar to the asymptotic analysis of the symbol error probability [21, 22,61]. Using [28, Eq. (8.445)] and [54, Eq. (4.60)], P_{md} can be given in an infinite series expression as

$$P_{md}(\gamma) = 1 - Q_N\left(\sqrt{2N\bar{\gamma}\beta}, \sqrt{\frac{\lambda}{\sigma_w^2}}\right) - \sum_{i=1}^{N-1}\sum_{n=0}^{\infty} \frac{N^n \bar{\gamma}^n \left(\frac{\lambda}{2\sigma_w^2}\right)^{i+n}}{n! e^{\frac{\lambda}{2\sigma_w^2}} \Gamma(i+n+1)} e^{-N\bar{\gamma}\beta} \beta^n. \tag{4.10}$$

The average missed-detection probability, $\overline{P_{md}}$, can be derived by integrating (4.10) over the PDF of channel-dependent random variable β, $f_\beta(\beta)$ in (4.9). It can be given as sum of a term of $\bar{\gamma}$ of order $-(t+1)$ and the term $\mathcal{O}(\bar{\gamma}^{-(t+1)})$. For high SNR, since $\overline{P_{md}}$ is dominated by the term of $\bar{\gamma}$ of order $-(t+1)$, it can be given as

$$\overline{P_{md}} \approx A\bar{\gamma}^{-t-1} \tag{4.11}$$

where A is a constant which is independent of $\bar{\gamma}$. Similar to the diversity order analysis in wireless communications [61], the slope of $\overline{P_{md}}$ versus $\bar{\gamma}$ curve (in log–log scale) shows how fast $\overline{P_{md}}$ decreases with the increase in the average SNR, which is given as $n_s = (t+1)$. Thus, sensing gain is defined as how fast the missed-detection probability decreases with the increase in SNR in the high SNR region.

Sensing gain indicates the detection performance under large SNR. A larger slope corresponds to a higher detection performance, and vice versa. Further, both Rayleigh and Rician fading channels have sensing gain of 1 ($t = 0$), and the Nakagami-m fading channel has sensing gain of m ($t = m-1$), which are confirmed by Fig. 4.1. Moreover, sensing gain can quantify the effect of diversity branches in cooperative spectrum sensing. In Fig. 4.2, $\overline{P_{md}}$ versus the average SNR is shown for the maximal ratio combining (MRC) over Rayleigh fading with the number of diversity branches being $L = 1, 2, 3$. The value L shows how fast the average missed-detection probability decreases with the increase in the SNR in the high SNR region, which means that sensing gain is L.

However, the sensing gain does not characterize the false alarm probability, and consequently does not represent the overall detection capability of a detector. Therefore, a comprehensive metric that includes both false-alarm and detection/missed-detection probabilities is needed. The ROC curve and AUC are two metrics that can show the trade-off between the false-alarm and detection/missed-detection probabilities, as follows.

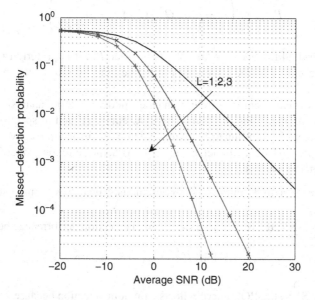

Fig. 4.2 The average missed-detection probability with average SNR of **S1** signal model for MRC over Rayleigh fading with $L = 1, 2, 3$, $2\sigma_w^2 = 1$, $N = 5$ and $\lambda = 5$

4.3   Receiver Operating Characteristic Curve

The ROC curve measures the sensitivity of a detector used in a binary classifier system [64]. The application of ROC curves has become much more popular over past years in the field of signal detection. It is often used to illustrate/visualize or to quantify the overall detection capability of the detector. In signal detection theory, the ROC (or the complementary ROC) curve is a graphical plot of P_d (or P_{md}) versus P_f as its discrimination threshold, λ, varies. Each point on the ROC curve represents a pair (P_f, P_d) for particular λ. In general, ROC curve can be given as

$$P_d = f(P_f). \tag{4.12}$$

It is difficult to express P_d in terms of P_f if exact analysis is used as in Sect. 2.3.2. In this case, ROC curve can be plotted by using pairs $(P_d(\lambda), P_f(\lambda))$ for particular threshold values λ. On the other hand, it is easy to express P_d in terms of P_f as in (4.12) if CLT approach is considered as in Sect. 2.3.3. From (2.19) and (2.20), λ can be eliminated, and ROC curve can be given for **S1** signal model as

$$P_d \approx Q\left(\frac{1}{\sqrt{1+2\gamma}}Q^{-1}(P_f) - \frac{\sqrt{N}\gamma}{\sqrt{1+2\gamma}}\right). \tag{4.13}$$

Since it is desired to have a higher P_d with lower P_f, it is preferred if the ROC curve can be moved to the upper left corner, i.e., the overall detection capability is high.

With the development of wireless channel modeling, ROC analysis of energy detector was an interesting problem. The ROC curves related to spectrum-sensing detectors have highly non-linear behavior, and they are, in general, convex. By using the ROC curve of **S1** signal model, given in (4.13), some common properties of ROC in energy detection are as follows

- The slope of a ROC curve, $\frac{dP_d}{dP_f}$, continuously increases from 0 to ∞ as λ increases;
- For a high SNR, the ROC curve shows good detection performance because

$$\lim_{\gamma \to \infty} P_d \to Q(-\infty) = 1;$$

- For a low SNR, the ROC curve reflects a random detection because

$$\lim_{\gamma \to 0} P_d \to Q\left(Q^{-1}(P_f)\right) = P_f.$$

Initially, the ROC has been analyzed for an unknown deterministic signal over a flat band-limited Gaussian noise channel. The first comprehensive ROC analysis of energy detection in wireless communications has been introduced in [58]. Since P_d increases with average SNR and P_f is independent of average SNR, the detection capability increases by increasing average SNR which is shown in Fig. 4.3 by using ROC curve.

In wireless communications, as P_d depends on the received instantaneous SNR, which is a function of the wireless channel gain, the average detection probability (or average missed-detection probability) over fading channels is important for plotting the ROC curve. Therefore, the work of Gaussian channel has been extended

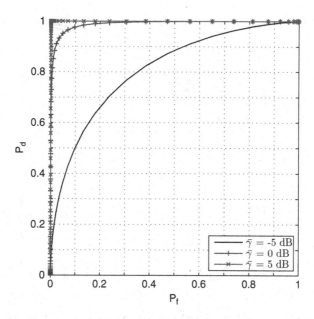

Fig. 4.3 The ROC curve of **S1** signal model over the Gaussian channel for different SNRs with $2\sigma_w^2 = 1$, $N = 20$ and $\lambda = 5$

to different other channels which include different multipath fading and shadowing effects. The results and techniques used in Sect. 4.1 would be helpful. It is shown that detection capability increases with the fading parameter m (e.g., Nakagami-m fading) and shadowing parameter k (e.g., Generalized-K_G fading), because as $m \to \infty$ and $k \to \infty$, the channel approaches the AWGN channel [7, 24, 35]. However, the rate of improvement diminishes as the increment of m or k. The overall detection can also be improved by using higher number of samples. Figure 4.4 shows that, for energy detection over Nakagami-m fading channels, ROC curves move to the upper left corner as increasing m or N, confirming better overall detection performance.

Furthermore, diversity reception schemes such as equal gain combining (EGC), selection combining (SC), MRC, square law combining (SLC), and square law selection (SLS) can boost the performance of the energy detector. The ROC curve reveals the diversity advantage, showing that MRC improves the performance of the energy detector the most among all the combining techniques [4, 7, 23, 24, 32–35]. The ROC analysis is also investigated for relay-based cognitive radio networks, and it is shown that the detection capability increases when the number of relay nodes increases. Furthermore, the direct path communication between the source and the destination has a major impact on the detection as introducing additional spatial diversity [5, 10]. When the average SNR, number of samples, number of diversity branches or number of relays increases, the ROC curve is shifted to the upper left corner of the ROC plot, which means better overall detection capability. These will be discussed in details in Chap. 5.

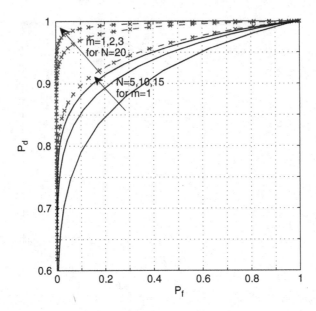

Fig. 4.4 The ROC curve of **S1** signal model for $\bar{\gamma} = 5\,\mathrm{dB}$ and $2\sigma_w^2 = 1$. *Solid lines* represent different N values, $N = 5, 10, 15$, for fixed $m = 1$ (Rayleigh fading), and *dashed lines* represent different m values, $m = 1, 2, 3$, for fixed $N = 20$

Although the ROC curve is conceptually simple, some properties which can be applied for research are not so obvious. Generally, the ROC curves are plotted by varying only one parameter (e.g., average SNR) while keeping other parameters fixed (e.g., number of samples, fading parameter). Therefore, a set of curves has to be generated for different combinations of parameters of interest. Although the ROC curves fully characterize the performance of an energy detector, it is difficult to compare performance of two energy detectors based on visual perception of their ROC curves, because the curves may cross each other. Even though the ROC curves show the performance increment due to spatial diversity (in diversity combining schemes or in relay networks), the order of increment is hidden. Therefore, it is desirable to have a single figure of merit to evaluate the overall detection capability. One such metric is the AUC.

4.4 Area Under the ROC Curve

The AUC is the area under the ROC curve which has been identified as a metric to represent different concepts of a hypothesis test, e.g.,

- A well-known theorem by Green, called Green's Theorem, has shown that the percentage of correctness is equal to the area under the Yes–No ROC curve [29].

- The area theorem proposed in [62] has shown that the AUC is a measure of the overall detection capability.
- In [30], it has shown that the area under the curve represents the probability that choosing the correct decision at the detector is more likely than choosing the incorrect decision.

The AUC is a compact and single value quantity, which is a measure for the overall detection capability of a detector. It varies between $\frac{1}{2}$ and 1. If the detector's performance is no better than flipping a coin, then the AUC becomes $\frac{1}{2}$, and it increases to one as the detector performance improves. As indicated in [19, 49], the exact computation of AUC is difficult for realistic detection tasks. Therefore, some research efforts, not necessarily for wireless communications, mainly focus on bounds, asymptotic expansions or limiting value of AUCs [17, 19, 49]. In [6], the AUC is rigorously analyzed for the energy detection of different scenarios in wireless communications. In Sects. 4.4.1–4.4.3, three methods to calculate AUC are discussed.

4.4.1   Direct Integration

The ROC curve can be given as $P_d = f(P_f)$ where P_d is a function of P_f for some cases. Then the AUC can be evaluated as

$$\mathscr{A} = \int_0^1 P_d \, dP_f. \tag{4.14}$$

For the ROC curve given in (4.13), the instantaneous AUC can be evaluated for **S1** signal model as

$$\mathscr{A}(\gamma) \approx 1 - Q\left(\frac{\sqrt{N}\gamma}{\sqrt{2(1+\gamma)}}\right). \tag{4.15}$$

The average AUC can be derived by averaging $\mathscr{A}$ over the SNR distribution $f_\gamma(x)$ as $\overline{\mathscr{A}} = \int_0^\infty \mathscr{A}(x) f_\gamma(x) dx$. The derivation of average AUC over fading channels is difficult for general case.

4.4.2   Threshold Averaging

When P_d cannot be written as a function of P_f (e.g., with exact expressions), (4.14) is not applicable. Then the threshold averaging method [27] can be used because both false alarm and detection probabilities are functions of λ, denoted as $P_f(\lambda)$ and $P_d(\lambda)$, respectively. Alternatively, the AUC in (4.14) can be given as

$$\mathscr{A} = \int_{\lambda \in \mathscr{R}} P_d(\lambda) \frac{\partial P_f(\lambda)}{\partial \lambda} d\lambda \tag{4.16}$$

where λ is in range $\mathscr{R}$ (real value set). For example, when the value of $P_f(\lambda)$ varies from $0 \rightarrow 1$, λ varies from $\infty \rightarrow 0$. For **S1** and **S2** signal models, instantaneous AUCs can be evaluated as [6]

$$\mathscr{A}_{S1}(\gamma) = 1 - e^{-\frac{N\gamma}{2}} \sum_{k=0}^{N-1} \frac{(N\gamma)^k}{2^k \Gamma(k+1)}$$

$$+ e^{-N\gamma} \sum_{r=1-N}^{N-1} \frac{\Gamma(N+r)_1 \tilde{F}_1\left(N+k; 1+k; \frac{N\gamma}{2}\right)}{2^{N+r} \Gamma(N)} \tag{4.17}$$

$$\mathscr{A}_{S2}(\gamma) = \frac{\Gamma(2N)(1+\gamma)^N {_2}\tilde{F}_1\left(N, 2N; 1+N; -(1+\gamma)\right)}{\Gamma(N)}$$

where $_1\tilde{F}_1(\cdot; \cdot; \cdot)$ and $_2\tilde{F}_1(\cdot; \cdot; \cdot)$ are regularized confluent hypergeometric function of the confluent hypergeometric function $_1F_1(\cdot; \cdot; \cdot)$ and $_2F_1(\cdot; \cdot; \cdot)$, respectively [63]. The average AUCs, $\overline{\mathscr{A}}$, under different fading channels for **S1** signal model can be derived as in [6], e.g., $\overline{\mathscr{A}}$ over Rayleigh fading is

$$\overline{\mathscr{A}} = 1 - \sum_{k=0}^{N-1} \frac{2(N\bar{\gamma})^k \Gamma(1+k)}{k!(2+N\bar{\gamma})^{k+1}} + \sum_{k=1-N}^{N-1} \frac{\Gamma(N+k)_2 \tilde{F}_1\left(1, k+N; 1+k; \frac{N\bar{\gamma}}{2(1+N\bar{\gamma})}\right)}{2^{N+k}(1+N\bar{\gamma})\Gamma(N)}. \tag{4.18}$$

Moreover, average AUCs for MRC, SLC, and SC diversity combining, with impacts of channel estimation errors, and with fading correlations are also discussed in [6]. For high SNR, the average AUC can be approximated as $\overline{A} \approx 1 - G\bar{\gamma}^{-n}$, where G is not a function of $\bar{\gamma}$ but a function of other system parameters such as N, fading parameter m, and number of diversity branches L, and n is a positive integer, e.g., for Nakagami-m fading, $n = m$ with no-diversity and $n = Lm$ with diversity (L is the number of branches). It is clearly seen that the average AUC approaches value 1 in order n which represents the detection diversity order. This will be discussed in details in Sect. 4.4.4.

4.4.3 MGF Approach

Using an alternative representation of P_d in (4.3), which uses a circular contour integral for the generalized Marcum-Q function, $\mathscr{A}$ can be written by using (4.16) as [11]

$$\mathscr{A}(\gamma) = \int_0^\infty \frac{e^{-\frac{\lambda}{2\sigma_w^2}}}{j2\pi} \oint_\Omega \frac{e^{(\frac{1}{z}-1)N\gamma + \frac{\lambda}{2\sigma_w^2}z}}{z^N(1-z)} dz \frac{\left(\frac{\lambda}{2\sigma_w^2}\right)^{N-1} e^{-\frac{\lambda}{2\sigma_w^2}}}{2\sigma_w^2 \Gamma(N)} d\lambda \tag{4.19}$$

$$= \frac{1}{j2\pi} \oint_\Omega \frac{e^{(\frac{1}{z}-1)N\gamma}}{z^N(1-z)(2-z)^N} dz.$$

Thus, the average AUC is

$$\overline{\mathscr{A}} = \frac{1}{j2\pi} \oint_\Omega \frac{\mathscr{M}_\gamma\left[\left(1-\frac{1}{z}\right)N\right]}{z^N(1-z)(2-z)^N} dz. \tag{4.20}$$

The average AUC over Rayleigh fading is derived as [11]

$$\overline{\mathscr{A}} = \frac{1}{(1+N\bar{\gamma})} \begin{cases} \left(\mathrm{Res}\left(g; \frac{N\bar{\gamma}}{1+N\bar{\gamma}}\right) + \mathrm{Res}\,(g;0)\right) : N > 1 \\ \mathrm{Res}\left(g; \frac{N\bar{\gamma}}{1+N\bar{\gamma}}\right) : \qquad\qquad N = 1 \end{cases} \tag{4.21}$$

where

$$g(z) = \frac{1}{z^{N-1}(z - \frac{N\bar{\gamma}}{1+N\bar{\gamma}})(1-z)(2-z)^N}.$$

Moreover, $\overline{\mathscr{A}}$ is also derived for Nakagami-m and η-μ fading channels [11].

The analytical approach of previous method in (4.18) may not suffice for some network scenarios (e.g., cooperative spectrum sensing), because the special functions (e.g., Marcum-Q, confluent hypergeometric, and regularized confluent hypergeometric functions) in the detection probability and the AUC expressions lead to high computational complexity. These drawbacks can be circumvented by this MGF approach, in which the residue calculations are simple, with no special functions involved.

On the other hand, by using the infinite series representation of P_d in (4.7) and the threshold averaging method, the instantaneous AUC is given as [1]

$$\mathscr{A}(\gamma) = \frac{1}{2^N N \Gamma(N)} \sum_{k=0}^\infty \frac{N^k \Gamma(k+2N)\,_2F_1(1, k+2N; 1+N; \frac{1}{2})}{k!2^{k+N}\Gamma(k+N)} \gamma^k e^{-N\gamma}. \tag{4.22}$$

Similarly to (4.8), the average AUC over a generalized fading channel, $f_\gamma(x)$, can be evaluated as

$$\overline{\mathscr{A}} = \frac{1}{2^N N \Gamma(N)} \sum_{k=0}^\infty \frac{(-1)^k N^k \Gamma(k+2N)\,_2F_1(1, k+2N; 1+N; \frac{1}{2})}{k!2^{k+N}\Gamma(k+N)} \mathscr{M}_\gamma^{(k)}(s)\big|_{s=N} \tag{4.23}$$

for which a series truncation is required for the numerical calculation.

As analyzed in [11], for energy detection over Nakagami-m fading channels, Fig. 4.5 shows the average AUC versus average SNR for different fading parameters m at $N = 20$ and for different N values at $m = 1$. AUCs of all cases vary from 0.5 to 1 as SNR increases from a very low value to a high value. A higher m or N leads to larger average AUC, and thus, higher overall detection capability.

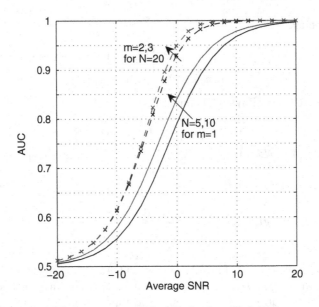

Fig. 4.5 The average AUC versus average SNR of **S1** signal model. *Solid lines* represent different N values, $N = 5, 10$, for fixed $m = 1$ (Rayleigh fading), and *dashed lines* represent different m values, $m = 2, 3$, for fixed $N = 20$

4.4.4 Complementary AUC

The multipath fading environment, diversity combining schemes, and relay networks introduce detection diversity gain on energy detection. However, neither the ROC curve nor the AUC curve is able to graphically show the order of improvement in detection capability clearly when the average SNR increases. Therefore, complementary AUC (CAUC) is introduced as a proxy for the overall detection capability, which can demonstrate the order of improvement when average SNR increases based on a log–log scale of a CAUC versus average SNR plot. The CAUC, $\mathscr{A}'$, which is the area under the complementary ROC curve (the curve of P_{md} versus P_f), is given as [8]

$$\mathscr{A}' = \int_0^1 (1 - P_d)\, dP_f = 1 - \mathscr{A}. \tag{4.24}$$

When the average SNR increases, while $\mathscr{A}$ goes from 0.5 to 1, $\mathscr{A}'$ goes from 0.5 to 0. By using the results of $\mathscr{A}$ given in previous sections, instantaneous CAUC and average CAUC, $\overline{\mathscr{A}'} = 1 - \overline{\mathscr{A}}$, can easily be evaluated.

Since $\mathscr{A}' \to 0$ as $\bar{\gamma} \to \infty$, CAUC is a better performance metric to discuss detection diversity order, which shows how fast the CAUC decreases with the increase in the SNR in the high SNR range. Thus, the detection diversity order can be defined as

$$d = -\lim_{\bar{\gamma} \to \infty} \frac{\log \mathscr{A}'}{\log \bar{\gamma}}. \tag{4.25}$$

In general, the average CAUC can be given as

$$\overline{\mathscr{A}'} = c\bar{\gamma}^{-d} + \mathcal{O}(\bar{\gamma}^{-(d+1)})$$

where c is a constant independent of $\bar{\gamma}$. For example, $\overline{\mathscr{A}'}$ over Nakagami-m fading channel is given as [6]

$$\overline{\mathscr{A}'} = \left(\frac{m}{N}\right)^m \left[\sum_{k=0}^{N-1} \frac{2^m \Gamma(k+m)}{k!\,\Gamma(m)} - \sum_{k=1-N}^{N-1} \frac{\Gamma(N+k)}{2^{N+k}\,\Gamma(N)} {}_2\tilde{F}_1\left(m, N+k; 1+k; \frac{1}{2}\right)\right]\bar{\gamma}^{-m}$$

$$+\mathcal{O}(\bar{\gamma}^{-(m+1)}).$$

$$\tag{4.26}$$

Thus, it is concluded that the energy detection achieves the detection diversity order of m over Nakagami-m fading channel, which is also shown in Fig. 4.6. If there are L diversity branches or K cooperative relays in a network, the detection diversity order is shown to be L or K, respectively, over Rayleigh fading channels. This will be discussed in details in Chap. 5.

4.4.5 Partial AUC

Although the AUC is a measure of the overall detection capability, it may not always unambiguously indicate when one detector is better than another. For example, when two ROC curves cross, it is possible that the AUC for the two ROC curves are the same. This situation can arise when the two associated detectors have different performance in different regions of detection threshold λ. The area of the ROC curve (for λ from 0 to ∞) only gives the overall detection performance, but cannot differentiate the two detectors in a partial region of λ, say $\lambda_1 \leq \lambda \leq \lambda_2$. To remedy this drawback, the partial area under the ROC curve, $\mathscr{A}_P$, in region (λ_1, λ_2) can be used to demonstrate the difference [40], as given by

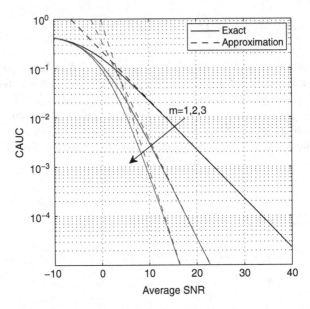

Fig. 4.6 The average CAUC versus average SNR of **S1** signal model for different *m* values (in Nakagami-*m* fading channel), $m = 1, 2, 3$, and $N = 10$. *Solid lines* represent exact values and *dashed lines* represent approximated values [the first term of (4.26)]

$$\mathscr{A}_P = \int_{\lambda_1}^{\lambda_2} P_d(\lambda) \frac{\partial P_f(\lambda)}{\partial \lambda} d\lambda. \qquad (4.27)$$

Nevertheless, the partial AUC measure appears intractable for closed-form analysis for energy detection. It can however be readily evaluated via numerical integration methods.

4.5   Low-SNR Energy Detection

Recently, the US FCC issued a report and order (R&O) permitting the TV white space spectrum to be used by fixed broadband access systems. The IEEE 802.22 WRAN standard also allows the use of the cognitive radio technique on a non-interfering basis [51, 55]. Since this basis is violated if a missed detection occurs at the spectrum-sensing device of a secondary user, both the FCC R&O and the IEEE 802.22 WRAN standard require reliable detection of primary user activities at a low SNR, e.g., SNR at −20 dB for a signal power of −116 dBm and a noise floor of −96 dBm. According to the IEEE 802.22, the allowable false alarm probability (which indicates the level of undetected spectrum holes) and missed-detection probability (which indicates the level of unexpected interference to primary users) should be both less than 0.1 with the receiver sensitivity being $-116\,\text{dBm}$ [20,50,55]. These two performance metrics reflect the overall efficiency and reliability of the cognitive network.

Although the energy detector performs well at moderate and high SNRs, where fine-sensing time can take in order of milliseconds (e.g., 25 ms), its performance is poor in low SNR. This problem is not limited to energy detection. For example, at a low SNR, a matched filter may lose the synchronization, and a cyclostationary detector may require a longer time to find a strong cyclostationary property. While increasing the sensing time is an option, IEEE 802.22 limits the maximal sensing time to 2 s. This maximal sensing time is useful at low SNR caused by severe fading, path loss, and shadowing [42,55]. These requirements (on false alarm probability, missed-detection probability, and sensing time) thus pose significant challenges for energy detection at a low SNR [65], which is a critically important topic in the wireless industry.

Under the low-SNR assumption (i.e., $\gamma \ll 1$), the signal has little impact on the variance of the test statistic under $\mathcal{H}_1$ [13, 14]. The distribution of test statistic is approximated as Gaussian which is given in (2.22), based on which the detection probability for any signal model (**S1, S2,** or **S3**) is given as

$$P_d \approx Q\left(\frac{\lambda - N(2\sigma_w^2)(1+\gamma)}{\sqrt{N}(2\sigma_w^2)}\right). \tag{4.28}$$

A basic analytical framework for the performance of energy detection at a low SNR is provided in [13] which develops a unified approach to evaluate the average detection probability and AUC based on a generalized channel model (i.e., mixture gamma distribution [9, 12]). For example, the average detection probability over the Rayleigh fading channel is

$$\overline{P_d} \approx 1 - \frac{1}{2}\left[\text{Erfc}\left(\frac{N\sigma_w^2 - \lambda}{\sqrt{2N}\sigma_w^2}\right) - e^{\frac{\frac{1}{\bar{\gamma}^2} + \frac{4}{\bar{\gamma}}\left(\frac{N\sigma_w^2 - \lambda}{\sqrt{2N\sigma_w^2}}\right)\sqrt{\frac{N}{2}}}{2N}}\,\text{Erfc}\left(\frac{N\sigma_w^2 - \lambda}{\sqrt{2N}\sigma_w^2} + \frac{1}{\bar{\gamma}\sqrt{2N}}\right)\right], \tag{4.29}$$

where $\text{Erfc}(\cdot)$ is the complementary error function defined as $\text{Erfc}(z) = \frac{2}{\sqrt{\pi}}\int_z^\infty e^{-t^2}\,dt$ [28]. The average AUC over a Rayleigh fading channel, $\overline{\mathscr{A}}$, is

$$\overline{\mathscr{A}} \approx \frac{1}{2} + \frac{e^{\frac{1}{N\bar{\gamma}^2}}}{2}\text{Erfc}\left(\frac{1}{\bar{\gamma}}\sqrt{\frac{1}{N}}\right). \tag{4.30}$$

Figure 4.7 shows the ROC curves for three fading scenarios: the AWGN, Rayleigh and Nakagami-4 fading channels with $-20\,\text{dB}$ average SNR. The ROC curves are plotted for $N = 2 \times 10^3$ and $N = 2 \times 10^5$. In the figure, the exact results are simulation results, represented by discrete marks, while analytical results are based on low-SNR approximation in (2.22), represented by solid lines. It can be seen that the analytical results perfectly match the simulation results for a high and low number of samples, and also for high and low P_d and P_f, confirming the accuracy of low-SNR approximation in (2.22). It can be seen that, when $N = 2 \times 10^3$, the IEEE 802.22 requirements on false alarm and missed-detection probabilities cannot be satisfied simultaneously in any case, and when $N = 2 \times 10^5$, the requirements can be satisfied simultaneously for the AWGN and Nakagami-4 fading channels but not for the Rayleigh fading case. Figure 4.7 clearly shows that a larger N improves detection performance.

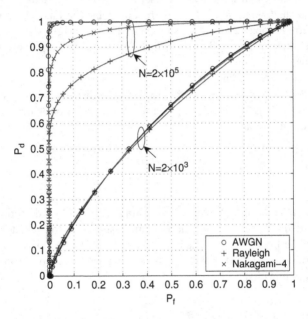

Fig. 4.7 The exact and approximated (low-SNR analysis) ROC curves of AWGN, Rayleigh and Nakagami-4 fading channels for $N = 2 \times 10^3$ and $N = 2 \times 10^5$ at -20 dB average SNR

In Fig. 4.7, the three ROC curves for $N = 2 \times 10^3$ intersect each other, making it difficult to compare the overall detection performance among the three fading scenarios. In such a case, the AUC, a single-valued measurement, is a better comparative performance metric. Table 4.1 shows the average AUC with different fading channels, in which the numbers outside brackets are analytical approximations and the numbers in brackets are exact AUC values (numerically calculated as the area under the curves in Fig. 4.7). It can be seen that approximations for AUC are accurate. As expected, AUC of energy detection over AWGN channel and that over Rayleigh channel vary from the largest to the smallest, and a larger number of samples leads to a higher AUC confirming a better overall detection capability.

Table 4.1 Exact (the numbers in brackets) and approximated (the number outside brackets) AUCs of AWGN, Rayleigh and Nakagami-4 channels in Fig. 4.7

Figure 4.7	$N = 2 \times 10^3$	$N = 2 \times 10^5$
AWGN	0.624085 (0.623489)	0.999217 (0.999174)
Nakagami-4	0.622408 (0.62172)	0.974777 (0.97462)
Rayleigh	0.616163 (0.615361)	0.895188 (0.895034)

References

1. Alam S., Odejide O., Olabiyi O., Annamalai A. (2011) Further results on area under the ROC curve of energy detectors over generalized fading channels. In: the 34th IEEE Sarnoff Symposium, Princeton, 3–4 May 2011.
2. Annamalai, A., Olabiyi, O., Alam, S., Odejide, O., Vaman, D. (2011) Unified analysis of energy detection of unknown signals over generalized fading channels. In: Proceedings of International Wireless Communications and Mobile Computing Conference (IWCMC), Istanbul, 4–8 July 2011.
3. Annamalai, A., Tellambura, C. (2008) A simple exponential integral representation of the generalized Marcum Q-function $Q_M(a, b)$ for real-order M with applications. In: Proceedings of IEEE Military Communications Conference (MILCOM), San Diego, 17–19 Nov 2008.
4. Atapattu, S., Tellambura, C., Jiang, H. (2009) Energy detection of primary signals over $\eta - \mu$ fading channels. In: Proceedings of International Conference Industrial and Information Systems (ICIIS), Kandy, 28–31 Dec 2009.
5. Atapattu, S., Tellambura, C., Jiang, H. (2009) Relay based cooperative spectrum sensing in cognitive radio networks. In: Proceedings of IEEE Global Telecommunications Conference (GLOBECOM), Hawaii, 30 Nov- 4 Dec 2009.
6. Atapattu, S., Tellambura, C., Jiang, H. (2010) Analysis of area under the ROC curve of energy detection. IEEE T on Wireless Communications 9(3): 1216–1225.
7. Atapattu, S., Tellambura, C., Jiang, H. (2010) Performance of an energy detector over channels with both multipath fading and shadowing. IEEE T on Wireless Communications 9(12): 3662–3670.
8. Atapattu, S., Tellambura, C., Jiang, H. (2010) Performance of energy detection: A complementary AUC approach. In: Proceedings of IEEE Global Telecommunications Conference (GLOBECOM), Miami, 6–10 Dec 2010.
9. Atapattu, S., Tellambura, C., Jiang, H. (2010) Representation of composite fading and shadowing distributions by using mixtures of gamma distributions. In: Proceedings of IEEE Wireless Communications and Networking Conference (WCNC), Sydney, 18–22 Apr 2010.
10. Atapattu, S., Tellambura, C., Jiang, H. (2011) Energy detection based cooperative spectrum sensing in cognitive radio networks. IEEE T on Wireless Communications 10(4): 1232–1241.
11. Atapattu, S., Tellambura, C., Jiang, H. (2011) MGF based analysis of area under the ROC curve in energy detection. IEEE Communications Letters 15(12): 1301–1303.
12. Atapattu, S., Tellambura, C., Jiang, H. (2011) A mixture gamma distribution to model the SNR of wireless channels. IEEE T on Wireless Communications 10(12): 4193–4203.
13. Atapattu, S., Tellambura, C., Jiang, H. (2011) Spectrum sensing via energy detector in low SNR. In: Proceedings of IEEE International Conference on Communications (ICC), Kyoto, 5–9 June 2011.
14. Atapattu, S., Tellambura, C., Jiang, H. (2011) Spectrum sensing in low SNR: Diversity combining and cooperative communications. In: Proceedings of International Conference Industrial and Information Systems (ICIIS), Kandy, 13–17 Aug 2011.

15. Banjade, V., Rajatheva, N., Tellambura, C. (2012) Performance analysis of energy detection with multiple correlated antenna cognitive radio in Nakagami-m fading. IEEE Communications Letters **16**(4): 502–505.
16. Banta, E. D. (1978) Energy detection of unknown deterministic signals in the presence of jamming. IEEE T on Aerospace and Electronic Systems **AES-14**(2): 384–386.
17. Barrett, H. H., Abbey, C. K., Clarkson, E. (1998) Objective assessment of image quality. III. ROC metrics, ideal observers, and likelihood-generating functions. J on Optical Society of America A **15**(6): 1520–1535.
18. Bello, P., Ehrman, L. (1969) Performance of an energy detection FSK digital modem for troposcatter links. IEEE T on Communications Technology **17**(2): 192–200.
19. Clarkson, E. (2002) Bounds on the area under the receiver operating characteristic curve for the ideal observer. J on Optical Society America A **19**(10): 1963–1968.
20. Cordeiro, C., Challapali, K., Birru, D., Shankar, S. N. (2006) IEEE 802.22: An introduction to the first wireless standard based on cognitive radios. J of Communications (JCM) **1(1)**: 38–47.
21. Dhungana, Y., Tellambura, C. (2012) New simple approximations for error probability and outage in fading. IEEE Communications Letters **16**(11): 1760–1763.
22. Dhungana, Y., Tellambura, C. (2013) Rational Gauss-Chebyshev quadratures for wireless performance analysis. IEEE Wireless Communications Letters **16**(2): 215–218.
23. Digham, F. F., Alouini, M. S., Simon, M. K. (2003) On the energy detection of unknown signals over fading channels. In: Proceedings of IEEE International Conference on Communications (ICC), Anchorage, 11–15 May 2003.
24. Digham, F. F., Alouini, M. S., Simon, M. K. (2007) On the energy detection of unknown signals over fading channels. IEEE T on Communications **55**(1): 21–24.
25. Dillard, G. M. (1973) Pulse-position modulation based on energy detection. IEEE T on Aerospace and Electronic Systems **AES-9**(4): 499–503.
26. Dillard, R. A. (1979) Detectability of spread-spectrum signals. IEEE T on Aerospace and Electronic Systems **AES-15**(4), 526–537.
27. Fawcett, T. (2006) An introduction to ROC analysis. Pattern Recognition Letters **27**(8): 861–874.
28. Gradshteyn, I. S., Ryzhik, I. M. (2000) Table of Integrals, Series, and Products, 6th edn, Academic Press, Inc.
29. Green, D. M. (1964) General prediction relating Yes-No and forced-choice results. J on Acoustics Society America A **36**(5): 1042–1042.
30. Hanley, J. A., Mcneil, B. J. (1982) The meaning and use of the area under a receiver operating characteristic (ROC) curve. Radiology **143**(1): 29–36.
31. Hauptschein, A., Knapp, T. (1979) Maximum likelihood energy detection of M-ary orthogonal signals. IEEE T on Aerospace and Electronic Systems **AES-15**(2): 292–299.
32. Herath, S. P., Rajatheva, N. (2008) Analysis of equal gain combining in energy detection for cognitive radio over Nakagami channels. In: Proceedings of IEEE Global Telecommunications Conference (GLOBECOM), New Orleans, 30 Nov-4 Dec 2008.
33. Herath, S. P., Rajatheva, N., Tellambura, C. (2009) On the energy detection of unknown deterministic signal over Nakagami channels with selection combining. In: Canadian Conference on Electrical and Computing Engineering (CCECE), Newfoundland, 3–6 May 2009.
34. Herath, S. P., Rajatheva, N., Tellambura, C. (2009) Unified approach for energy detection of unknown deterministic signal in cognitive radio over fading channels. In: Proceedings of IEEE International Conference on Communications (ICC) Workshops, Dresden, 14–18 June 2009.
35. Herath, S. P., Rajatheva, N., Tellambura, C. (2011) Energy detection of unknown signals in fading and diversity reception. IEEE T on Communications **59**(9): 2443–2453.
36. Karmeshu, Agrawal, R. (2007) On efficacy of Rayleigh-inverse Gaussian distribution over K-distribution for wireless fading channels. Wireless Communications and Mobile Computing **7**(1): 1–7.
37. Kostylev, V. I. (2002) Energy detection of a signal with random amplitude. In: Proceedings of IEEE International Conference on Communications (ICC), New York City, 28 Apr-2 May 2002.

38. Krantz, S. G. (1999) Handbook of Complex Variables, 1st Ed, Birkhäuser Boston.
39. Li, R., Kam, P. Y., Fu, Y. (2010) New representations and bounds for the generalized Marcum Q-function via a geometric approach, and an application. IEEE T on Communications **58**(1): 157–169.
40. Liu, A., Schisterman, E. F., Wu, C. (2005) Nonparametric estimation and hypothesis testing on the partial area under receiver operating characteristic curves. Communications in Statistics - Theory and Methods **34**(9): 2077–2088.
41. Mihos, S. K., Kapinas, V. M., Karagiannidis, G. K. (2008) Lower and upper bounds for the generalized Marcum and Nuttall Q-functions. In: Proceedings of International Symposium on Wireless and Pervasive Computing (ISWPC), Santorini, 7–9 May 2008.
42. Min, A. W., Shin, K. G., Hu, X. (2011) Secure cooperative sensing in IEEE 802.22 WRANs using shadow fading correlation. IEEE T on Mobile Computing **10**(10): 1434–1447.
43. Molisch, A. F., Greenstein, L. J., Shafi, M. (2009) Propagation issues for cognitive radio. Proceedings of the IEEE **97**(5): 787–804.
44. Nuttall, A. H. (1972) Some integrals involving the Q-function. Naval underwater Systems Center (NUSC) technical report.
45. Nuttall, A. H. (1974) Some integrals involving the Q_M-function. Naval underwater Systems Center (NUSC) technical report.
46. Olabiyi, O., Annamalai, A. (2011) Further results on the performance of energy detector over generalized fading channels. In: IEEE International Symposium on Personal, Indoor and Mobile Radio Communications (PIMRC), Toronto, 11–14 Sept 2011.
47. Pandharipande, A., Linnartz, J. P. M. G. (2007) Performance analysis of primary user detection in a multiple antenna cognitive radio. In: Proceedings of IEEE International Conference on Communications (ICC), Glasgow, 24–28 June 2007.
48. Park, K. Y. (1978) Performance evaluation of energy detectors. IEEE T on Aerospace and Electronic Systems **AES-14**(2): 237–241.
49. Shapiro, J. H. (1999) Bounds on the area under the ROC curve. J on Optical Society America A **16**(1): 53–57.
50. Shellhammer, S. J. (2008) Spectrum sensing in IEEE 802.22. In: 1st IAPR Workshop on Cognitive Information Processing, Santorini (Thera), 9–10 June 2008.
51. Shellhammer, S. J., Sadek, A. K., Zhang, W. (2009) Technical challenges for cognitive radio in the TV white space spectrum. In: Proceedings of Information Theory and Applications (ITA) Workshop, California, 8–13 Feb 2009.
52. Simon, M. K. (2002) The Nuttall Q function - its relation to the Marcum Q function and its application in digital communication performance evaluation. IEEE T on Communications **50**(11): 1712–1715.
53. Simon, M. K., Alouini, M. S. (2000) Exponential-type bounds on the generalized Marcum Q-function with application to error probability analysis over fading channels. IEEE T on Communications **48**(3): 359–366.
54. Simon, M. K., Alouini, M. S. (2005) Digital Communication over Fading Channels 2nd Ed, New York: Wiley.
55. Stevenson, C., Chouinard, G., Lei, Z., Hu, W., Shellhammer, S. J., Caldwell, W. (2009) IEEE 802.22: The first cognitive radio wireless regional area network standard. IEEE Communications M **47**(1): 130–138.
56. Sun, H., Laurenson, D., Wang, C. X. (2010) Computationally tractable model of energy detection performance over slow fading channels. IEEE Communications Letters **14**(10): 924–926.
57. Tellambura, C., Annamalai, A., Bhargava, V. K. (2003) Closed form and infinite series solutions for the MGF of a dual-diversity selection combiner output in bivariate Nakagami fading. IEEE T on Communications **51**(4): 539–542.
58. Urkowitz, H. (1967) Energy detection of unknown deterministic signals. Proceedings of the IEEE **55**(4): 523–531.
59. Urkowitz, H. (1969) Energy detection of a random process in colored Gaussian noise. IEEE T on Aerospace and Electronic Systems **AES-5**(2): 156–162.

60. Wang, Q., Yue, D. W. (2009) A general parameterization quantifying performance in energy detection. IEEE Signal Processing Letters **16**(8): 699–702.
61. Wang, Z., Giannakis, G. B. (2003) A simple and general parameterization quantifying performance in fading channels. IEEE T on Communications **51**(8): 1389–1398.
62. Wickens, T. D. (2002) Elementary Signal Detection Theory, New York: Oxford Univ. Press.
63. Wolfram Research. The Wolfram functions site: http://functions.wolfram.com.
64. Zhang, J., Mueller, S. T. (2005) A note on ROC analysis and non-parametric estimate of sensitivity. Psychometrika **70**(1): 203–212.
65. Zhang, W., Sadek, A. K., Shen, C., Shellhammer, S. J. (2010) Adaptive spectrum sensing. In: Proceedings of Information Theory and Applications (ITA) Workshop, San Diego, 31 Jan- 5 Feb 2010.

Chapter 5
Diversity Techniques and Cooperative Networks

Achieving IEEE 802.22 WRAN spectrum sensing specifications is challenging because of shadowing, fading, and time variation of wireless channels. These distortions can be mitigated by using diversity combining and cooperative spectrum sensing techniques. The former improves the receive SNR, and the latter mitigates the hidden terminal problem.

5.1 Traditional Diversity Techniques

Traditional diversity combining techniques such as MRC, EGC and SC improve the receive SNR. But they may increase the implementation complexity because the receiver may require additional knowledge of channel state information (CSI). Although some requirements may not be applicable with the notion of energy detection, these techniques can give the reference of achievable performance which can be used to compare with other practical settings.

The energy detector processes the samples of the combined signal of L diversity branches, $\mathbf{Y}(n)$, which can be given as

$$\mathbf{Y}(n) = \mathbf{H}s(n) + \mathbf{W}(n) \tag{5.1}$$

where $\mathbf{H}$ and $\mathbf{W}(n)$ are effective channel gain and noise sample, respectively. Thus, the test statistic is given as

$$\Lambda = \sum_{n=1}^{N} |\mathbf{Y}(n)|^2. \tag{5.2}$$

The effective number of samples for the test statistic is N.

S. Atapattu et al., *Energy Detection for Spectrum Sensing in Cognitive Radio*,
SpringerBriefs in Computer Science, DOI 10.1007/978-1-4939-0494-5_5,
© The Author(s) 2014

5.1.1   Maximal Ratio Combining

The MRC is a coherent combining technique which needs CSI in non-coherent energy detection. Thus, it may increase the design complexity. However, in cognitive radio applications, the CSI may be available to secondary users over a control channel or over a broadcast channel through an access point.

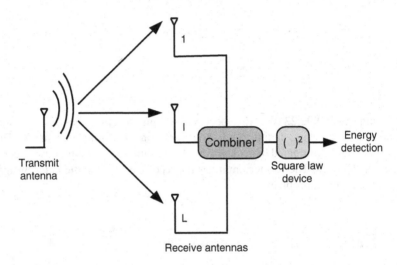

Fig. 5.1 Energy detection with MRC

The MRC receiver combines all the diversity branches weighted with their corresponding complex fading gains as shown in Fig. 5.1. The instantaneous SNR of the combiner output is thus the sum of the SNRs on all L branches. The output signal of the MRC can be written as (5.1) with

$$\mathbf{H} = \sum_{l=1}^{L} |\mathbf{h}_l|^2, \quad \mathbf{W}(n) = \sum_{l=1}^{L} \mathbf{h}_l^* \mathbf{w}_l(n),$$

where $\mathbf{w}_l(n)$ and $\mathbf{h}_l$ are the noise and channel coefficient of the lth branch, respectively, and $\mathbf{h}_l^*$ is the complex conjugate of $\mathbf{h}_l$. For **S1** signal model, the output signal has following distributions under two hypotheses:

$$\mathbf{Y}(n) = \begin{cases} \mathcal{CN}\left(0, 2\sigma_w^2 \sum_{l=1}^{L} |\mathbf{h}_l|^2\right) & : \mathcal{H}_0 \\ \mathcal{CN}\left(\mathbf{s}(n) \sum_{l=1}^{L} |\mathbf{h}_l|^2, 2\sigma_w^2 \sum_{l=1}^{L} |\mathbf{h}_l|^2\right) & : \mathcal{H}_1. \end{cases} \tag{5.3}$$

The test statistic of MRC, Λ_{MRC}, under $\mathcal{H}_0$ is a sum of $2N$ squares of independent Gaussian random variables with zero mean, and thus it follows central chi-square distribution. The exact false-alarm probability, $P_{f,MRC}$, can be derived as

$$P_{f,MRC} = \frac{\Gamma\left(N, \frac{\lambda}{2\sigma_w^2 \sum_{l=1}^{L} |\mathbf{h}_l|^2}\right)}{\Gamma(N)}. \tag{5.4}$$

Under $\mathscr{H}_1$, Λ_{MRC} is a sum of $2N$ squares of independent Gaussian random variables with non-zero mean, and thus it follows a non-central chi-square distribution. The exact detection probability, $P_{d,MRC}$, can be derived as

$$P_{d,MRC} = Q_N\left(\sqrt{2N\gamma_{MRC}}, \sqrt{\frac{\lambda}{\sigma_w^2 \sum_{l=1}^{L} |\mathbf{h}_l|^2}}\right) \tag{5.5}$$

where $\gamma_{MRC} = \sum_{l=1}^{L} \gamma_l$ and $\gamma_l = \frac{|\mathbf{h}_l|^2 \frac{1}{N}\sum_{n=1}^{N} |s(n)|^2}{2\sigma_w^2}$ which is the instantaneous SNR of the lth branch. As it can be seen from (5.4) and (5.5), both instantaneous $P_{f,MRC}$ and $P_{d,MRC}$ depend on channel gains, i.e., $\sum_{l=1}^{L} |\mathbf{h}_l|^2$. Since the receiver has knowledge of CSIs, the test statistics can be defined as $\Lambda_{MRC} = \frac{1}{\sum_{l=1}^{L} |\mathbf{h}_l|^2} \sum_{n=1}^{N} |\mathbf{Y}(n)|^2$. Then instantaneous false-alarm and detection probabilities can be given as

$$P_{f,MRC} = \frac{\Gamma\left(N, \frac{\lambda}{2\sigma_w^2}\right)}{\Gamma(N)}, \quad P_{d,MRC} = Q_N\left(\sqrt{2N\gamma_{MRC}}, \sqrt{\frac{\lambda}{\sigma_w^2}}\right). \tag{5.6}$$

The average detection probability can be derived by averaging $P_{d,MRC}$ over the SNR distribution of γ_{MRC} as $\overline{P_{d,MRC}} = \int_0^\infty P_{d,MRC}(x) f_{\gamma_{MRC}}(x)\, dx$.

5.1.2 Equal Gain Combining

While the MRC requires full channel knowledge (i.e., both channel amplitude and phase) of all diversity branches, the EGC requires only the knowledge of phase which offers a reduced complexity. The received signal via each diversity branch is weighted with the phase where ϕ_l is the phase of the lth diversity branch. The output signal of the EGC can be written as (5.1) with

$$\mathbf{H} = \sum_{l=1}^{L} |\mathbf{h}_l|, \quad \mathbf{W}(n) = \sum_{l=1}^{L} \mathbf{w}_l(n)e^{-j\phi_l}.$$

For **S1** signal model, the output signal has following distributions under two hypotheses for given channels:

$$\mathbf{Y}(n) = \begin{cases} \mathscr{C}\mathscr{N}\left(0, 2\sigma_w^2 L\right) & : \mathscr{H}_0 \\ \mathscr{C}\mathscr{N}\left(s(n)\sum_{l=1}^{L} |\mathbf{h}_l|, 2\sigma_w^2 L\right) & : \mathscr{H}_1. \end{cases} \tag{5.7}$$

Similar to the MRC case, the test statistic of EGC, Λ_{EGC}, follows a central chi-square or a non-central chi-square distribution under $\mathscr{H}_0$ or under $\mathscr{H}_1$, respectively. The exact false-alarm probability, $P_{f,EGC}$, and detection probability, $P_{d,EGC}$ can be derived as

$$P_{f,EGC} = \frac{\Gamma\left(N, \frac{\lambda}{2\sigma_w^2 L}\right)}{\Gamma(N)}, \quad P_{d,EGC} = Q_N\left(\sqrt{2N\gamma_{EGC}}, \sqrt{\frac{\lambda}{\sigma_w^2 L}}\right) \qquad (5.8)$$

respectively, where $\gamma_{EGC} = \frac{\left(\sum_{l=1}^{L}|\mathbf{h}_l|\right)^2 \frac{1}{N}\sum_{n=1}^{N}|\mathbf{s}(n)|^2}{L\sigma_w^2}$. If the test statistic is defined as $\Lambda_{EGC} = \frac{1}{L}\sum_{n=1}^{N}|\mathbf{Y}(n)|^2$, instantaneous false-alarm and detection probabilities can be given as

$$P_{f,EGC} = \frac{\Gamma\left(N, \frac{\lambda}{2\sigma_w^2}\right)}{\Gamma(N)}, \quad P_{d,EGC} = Q_N\left(\sqrt{2N\gamma_{EGC}}, \sqrt{\frac{\lambda}{\sigma_w^2}}\right). \qquad (5.9)$$

The average detection probability can be derived by averaging $P_{d,EGC}$ over the SNR distribution of γ_{EGC} as $\overline{P_{d,EGC}} = \int_0^\infty P_{d,EGC}(x) f_{\gamma_{EGC}}(x)\,dx$.

Similar to the case without diversity reception, performance analysis of an energy detector with traditional diversity reception can be discussed by using the performance metrics explained in Chap. 4. Energy detection with MRC, EGC, SC or switch-and-stay combining has been analyzed over Rayleigh, Nakagami-m, Rician, or K-Channel fading in [6–8, 16, 17], in which the average detection probability or average AUC is derived by using direct averaging, MGF approach, or infinite series representation.

5.2 Square-Law Techniques

In the non-coherent energy detection, having the instantaneous individual branch energy measurements along with the CSI at the receiver is infeasible. Hence, the non-coherent combining schemes which exploit the diversity gain in the absence of CSI are more preferable.

5.2.1 Square-Law Combining

In contrast to the MRC, each diversity branch in SLC has a square-law device which performs the square-and-integrate operation. The combiner is implemented following the square-law operation. The energy detector receives the sum of L

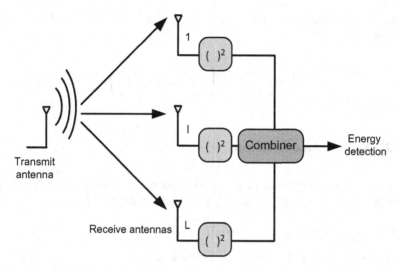

Fig. 5.2 Energy detection with SLC

decision statistics as shown in Fig. 5.2. The outputs of the square-law devices of L branches are combined to yield a new decision statistic as

$$\Lambda_{SLC} = \sum_{i=1}^{L} \Lambda_i = \sum_{i=1}^{L} \sum_{n=1}^{N} |\mathbf{y}_i(n)|^2$$

where Λ_i is the test statistic of the ith branch and $\mathbf{y}_i$ is received signal via the ith branch.

Under $\mathcal{H}_0$, Λ_{SLC} is a sum of $2LN$ squares of i.i.d. Gaussian random variables which follow $\mathcal{N}(0, \sigma_w^2)$. Thus Λ_{SLC} follows a central chi-square distribution with $2LN$ degree of freedom, and the false alarm probability can be given as

$$P_{f,SLC} = \frac{\Gamma(LN, \frac{\lambda}{2\sigma_w^2})}{\Gamma(LN)}. \tag{5.10}$$

Under $\mathcal{H}_1$, Λ_{SLC} is a sum of $2LN$ squares of independent and non-identically distributed Gaussian random variables, in which there are LN number of $\mathcal{N}(h_{lr}s_r(n) - h_{li}s_i(n), \sigma_w^2)$ random variables and LN number of $\mathcal{N}(h_{lr}s_i(n) + h_{li}s_r(n), \sigma_w^2)$ random variables. Here h_{lr} and h_{li} are real and imaginary component of $\mathbf{h}_l$. Thus, Λ_{SLC} follows a non-central chi-square distribution with $2LN$ degree of freedom and non-centrality parameter

$$\mu = 2N \frac{\sum_{l=1}^{L} |\mathbf{h}_l|^2 \frac{1}{N} \sum_{n=1}^{N} |\mathbf{s}(n)|^2}{2\sigma_w^2} = 2N\gamma_{SLC}$$

where γ_{SLC} has the same expression as γ_{MRC}. The detection probability can be given as

$$P_{d,SLC} = Q_{LN}\left(\sqrt{2N\gamma_{SLC}}, \sqrt{\frac{\lambda}{\sigma_w^2}}\right). \tag{5.11}$$

5.2.2   Square-Law Selection

In SLS, each diversity branch has a square-law device, and the branch with the maximum decision statistic is selected. Thus, the test statistic of SLS is

$$\Lambda_{SLS} = \max\{\Lambda_1, \cdots, \Lambda_l, \cdots, \Lambda_L\}.$$

Since Λ_l's are i.i.d., the CDF of Λ_{SLS} can be written as

$$F_{\Lambda_{SLS}}(x) = \Pr[\Lambda_{SLS} \le x] = \Pr[\Lambda_1 \le x, \cdots, \Lambda_l \le x, \cdots, \Lambda_L \le x] = \prod_{l=1}^{L} F_{\Lambda_l}(x)$$

where $F_{\Lambda_l}(x)$ is the CDF of Λ_l. Thus, the false alarm probability can be given as

$$P_{f,SLS} = \Pr[\Lambda_{SLS} \ge \lambda|\mathcal{H}_0] = 1 - \Pr[\Lambda_{SLS} \le \lambda|\mathcal{H}_0] = 1 - \prod_{l=1}^{L} F_{\Lambda_l|\mathcal{H}_0}(\lambda).$$

Let P_f given in (2.11) denote the false alarm probability of a branch. Then, $P_{f,SLS}$ can be given as

$$P_{f,SLS} = 1 - \left(1 - P_f\right)^L. \tag{5.12}$$

Similarly, the detection probability can be given as

$$P_{d,SLS} = 1 - \prod_{l=1}^{L}\left(1 - P_{d_l}\right) \tag{5.13}$$

where P_{d_l} is the lth branch detection probability which can be given as $P_{d_l} = Q_N\left(\sqrt{2N\gamma_l}, \frac{\sqrt{\lambda}}{\sigma_w}\right)$ as given in (2.13).

5.2.3   Performance Analysis

The average detection probabilities of SLC, $\overline{P_{d,SLC}}$, and SLS, $\overline{P_{d,SLS}}$, can be derived by averaging their instantaneous detection probabilities over their SNR PDFs.

For SLC with i.i.d. branches, we have $\overline{P_{d,SLC}} = \int_0^\infty P_{d,SLC}(x) f_{\gamma_{SLC}}(x)\,dx$ where $f_{\gamma_{SLC}}(x)$ is the PDF of γ_{SLC} which can be given for Rayleigh fading channels as

$$f_{\gamma_{SLC}}(x) = \frac{x^{L-1} e^{-\frac{x}{\bar{\gamma}_l}}}{(L-1)! \bar{\gamma}_l^L}$$

where $\bar{\gamma}_l$ is the average SNR of the lth branch, and it has the same form as gamma distribution. Thus, $\overline{P_{d,SLC}}$ has the same form as the average detection probability over Nakagami-m fading channels with no diversity, and can be evaluated by using any technique given in Sect. 4.1 [9, 16]. Similarly, average AUC can be evaluated by using any technique given in Sect. 4.4.

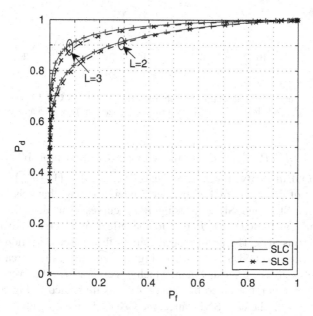

Fig. 5.3 The ROC curve of **S1** signal model over Rayleigh fading channels for SLC and SLS with $2\sigma_w^2 = 1$, $N = 10$ and $\bar{\gamma} = 0\,\mathrm{dB}$

For SLS, as γ_l's are independent, $\overline{P_{d,SLS}}$ can be derived as

$$\overline{P_{d,SLS}} = 1 - \int_0^\infty \cdots \int_0^\infty \prod_{l=1}^L (1-P_{d_l}) f_{\gamma_l}(x_l)\,dx_l = 1 - \prod_{l=1}^L \int_0^\infty (1-P_{d_l}) f_{\gamma_l}(x)\,dx$$

$$= 1 - \prod_{l=1}^L (1 - \overline{P_{d_l}})$$

$$(5.14)$$

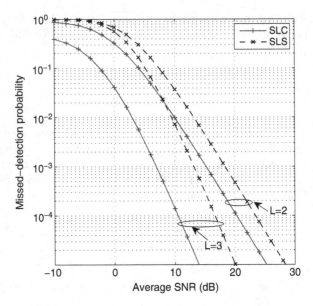

Fig. 5.4 The average missed-detection probability with average SNR of **S1** signal model over Rayleigh fading channels for SLC and SLS with $L = 2, 3, 2\sigma_w^2 = 1$, $N = 5$ and $\lambda = 15$

where $\overline{P_{d_l}} = \int_0^\infty P_{d_l} f_{\gamma_l}(x)\, dx$ is the average detection probability of the lth branch. For i.i.d. diversity branches, we have $\overline{P_{d,SLS}} = 1 - \left(1 - \overline{P_{d_l}}\right)^L$.

The effect of diversity combining over Rayleigh fading channels is illustrated in Fig. 5.3 for SLC and SLS by using ROC curves. The detection capability is significantly increased with L due to the effect of diversity advantage, and SLC outperforms SLS. The performance gain on the missed-detection probability achieved by SLC over SLS is shown in Fig. 5.4, and the gain increases with L. Further, Fig. 5.4 gives the sensing gain, which has order of L. The overall detection probability which is represented by using CAUC is illustrated in Fig. 5.5 for SLC with $L = 2, 3, 4$. In the high SNR range, the CAUC decreases with the increase in the SNR with the order of L, which confirms that the detection diversity order is L, as explained in Sect. 4.4.4.

5.3 Cooperative Networks

The hidden terminal problem, which occurs when the link from a primary transmitter to a secondary user is shadowed (e.g., there is a tall building between them) while a primary receiver is operating in the vicinity of the secondary user, presents a tough challenge. The secondary user may fail to notice the presence of the primary transmitter, and then accesses the licensed channel and causes interference

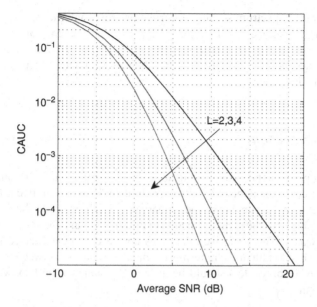

Fig. 5.5 The average CAUC with average SNR of **S1** signal model over Rayleigh fading channels for SLC with $L = 2, 3, 4$ and $N = 10$

to the primary receiver. In order to deal with the hidden terminal problem in cognitive radio networks, cooperative spectrum sensing has been introduced, in which single cognitive relay or multiple cognitive relays are introduced to the secondary network, as shown in Fig. 5.6. Cognitive relays individually sense the spectrum, and send their collected data to a fusion center. Since the cognitive relays are normally scattered in the network, the effect of the hidden terminal problem can be largely reduced.

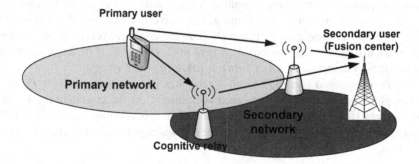

Fig. 5.6 Cooperative spectrum sensing in a cognitive radio network

In cooperative spectrum sensing, information from multiple cognitive relays is combined at the fusion center to make a decision on the presence or absence of the primary user. When energy detection is utilized for cooperative spectrum sensing, cognitive relays report to the fusion center their sensing data, in either data fusion or decision fusion mode.

5.3.1   Data Fusion

A simple cooperative spectrum sensing network has a primary user, a cognitive relay and a fusion center as depicted in Fig. 5.7. In data fusion, cognitive relay simply amplifies the received signal from the primary user and forwards to the fusion center [10,11,23], which has the similar function to the amplify-and-forward (AF) relaying in relay networks. The cognitive relay does not need complex detection process. However, the bandwidth of the reporting channel (the channel between the cognitive relay and the fusion center) should be at least the same as the bandwidth of the sensed channel.

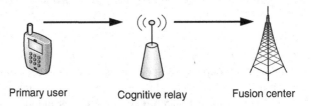

Primary user   Cognitive relay   Fusion center

Fig. 5.7 A cooperative spectrum sensing network by using cognitive relay

To model cooperative spectrum sensing, the complex-valued fading coefficients from the primary user to the cognitive relay and from the cognitive relay to the fusion center are denoted as $\mathbf{f}$ and $\mathbf{g}$, respectively, which are constant within each spectrum sensing period. The AWGN at the cognitive relay and the fusion center are $\mathbf{u}(n)$ and $\mathbf{v}(n)$, respectively, where $\mathbf{u}(n), \mathbf{v}(n) \sim \mathscr{CN}(0, 2\sigma_w^2)$, and their real/imaginary components $u_r(n), u_i(n), v_r(n), v_i(n) \sim \mathscr{N}(0, \sigma_w^2)$.

It can be assumed that the cognitive relay has its own power budget P_r, and the amplification factor, α, is designed accordingly. At the cognitive relay, first the received signal power is normalized, and then it is amplified by P_r. The CSI requirement depends on the relaying strategy, in which there are two types of relays [4, 15, 18]:

- Non-coherent power coefficient: the relay has knowledge of the average fading power of the channel between the primary user and itself, i.e., $\mathbb{E}[|\mathbf{f}|^2]$, and uses it to constrain its average transmit power. Therefore, α is given as

$$\alpha = \frac{1}{2\sigma_w^2 + P_s \mathbb{E}[|\mathbf{f}|^2]} \qquad (5.15)$$

where P_s is the transmit power at the source.

- Coherent power coefficient: the relay has knowledge of the instantaneous CSI of the channel between the primary user and itself, i.e., $\mathbf{f}$, and uses it to constrain its transmit power. Therefore, α is given as

$$\alpha = \frac{1}{2\sigma_w^2 + P_s |\mathbf{f}|^2}. \qquad (5.16)$$

An advantage of the non-coherent power coefficient over the coherent one is in its less overhead, because it does not need the instantaneous CSI, which requires training and channel estimation at the relay.

The fusion center implements with an energy detector, which decides whether a signal is present or not by using the received signal, $\mathbf{y}$. The decision is made from a binary hypothesis. In general, for hypothesis $\mathcal{H}_\theta$, $\theta \in \{0, 1\}$, it can be given as

$$\mathbf{y}(n) = \theta \sqrt{P_r \alpha} \mathbf{fgs}(n) + \sqrt{P_r \alpha} \mathbf{gu}(n) + \mathbf{v}(n). \qquad (5.17)$$

5.3.1.1 Distributions of Test Statistic

Under $\mathcal{H}_0$ (i.e., $\theta = 0$), $\mathbf{y}(n)$ depends on $\mathbf{g}$, $\mathbf{u}(n)$, and $\mathbf{v}(n)$. For a given $\mathbf{g}$, $\mathbf{y}(n)$ follows as

$$\mathbf{y}(n)|\mathcal{H}_0 \sim \mathcal{CN}\left(0, 2\Sigma_w^2\right)$$

where $\Sigma_w^2 = (P_r \alpha |\mathbf{g}|^2 + 1)\sigma_w^2$, and $y_r(n)$, $y_i(n) \sim \mathcal{N}\left(0, \Sigma_w^2\right)$. Since the test statistic Λ is a sum of $2N$ squares of independent random variables, its distribution $f_{\Lambda|\mathcal{H}_0}(x)$ follows central chi-square distribution which is given as [21]

$$f_{\Lambda|\mathcal{H}_0}(x) = \frac{x^{N-1} e^{-\frac{x}{2\Sigma_w^2}}}{\left(2\Sigma_w^2\right)^N \Gamma(N)}. \qquad (5.18)$$

Under $\mathcal{H}_1$, $\mathbf{y}(n)$ depends on $s(n)$, $\mathbf{f}$, $\mathbf{g}$, $\mathbf{u}(n)$, and $\mathbf{v}(n)$, and thus for given $\mathbf{f}$ and $\mathbf{g}$, $\mathbf{y}(n)$ has different distributions for different models for the signal ($\mathbf{S1}$, $\mathbf{S2}$ and $\mathbf{S3}$).

- $\mathbf{S1}$: the signal to be detected is Gaussian with non-zero mean, i.e., $\mathbb{E}[\mathbf{y}(n)] = \sqrt{P_r \alpha} \mathbf{fgs}(n)$, and thus

$$\mathbf{y}(n)|\mathbf{S1} \sim \mathcal{CN}\left(\sqrt{P_r \alpha} \mathbf{fgs}(n), 2\Sigma_w^2\right).$$

Here, $y_r(n)$ and $y_i(n)$ follow $\mathcal{N}\left(\mu_r, \Sigma_w^2\right)$ and $\mathcal{N}\left(\mu_i, \Sigma_w^2\right)$, respectively, where

$$\mu_r = \sqrt{P_r}\alpha(f_r g_r s_r(n) - f_r g_i s_i(n) - f_i g_i s_r(n) - f_i g_r s_i(n))$$

and

$$\mu_i = \sqrt{P_r}\alpha(f_r g_r s_i(n) + f_r g_i s_r(n) + f_i g_r s_r(n) - f_i g_i s_i(n)).$$

Since Λ is a sum of $2N$ squares of independent and non-identically distributed Gaussian random variables with non-zero mean, Λ follows non-central chi-square distribution with PDF given as

$$f_{\Lambda|\mathcal{H}_1,\mathbf{S1}}(x) = \frac{\left(\frac{x}{\Sigma_w^2}\right)^{\frac{N-1}{2}} e^{-\frac{1}{2}\left(\frac{x}{\Sigma_w^2}+\mu\right)} I_{N-1}\left(\sqrt{\frac{\mu x}{\Sigma_w^2}}\right)}{2\Sigma_w^2 \mu^{\frac{N-1}{2}}} \tag{5.19}$$

where μ is the non-centrality parameter

$$\mu = \sum_{n=1}^{N}\left[\frac{\mathbb{E}[y_r(n)]^2}{\mathbb{V}\mathrm{ar}[y_r(n)]} + \frac{\mathbb{E}[y_i(n)]^2}{\mathbb{V}\mathrm{ar}[y_i(n)]}\right] = \frac{P_r\alpha|\mathbf{f}|^2|\mathbf{g}|^2 \sum_{n=1}^{N}|s(n)|^2}{\Sigma_w^2} = 2N\gamma_e$$

where $\gamma_e = \frac{P_r\alpha|\mathbf{f}|^2|\mathbf{g}|^2 \frac{1}{N}\sum_{n=1}^{N}|s(n)|^2}{2\Sigma_w^2}$ is the end-to-end SNR, or in general, $\gamma_e = \frac{P_r\alpha|\mathbf{f}|^2|\mathbf{g}|^2 P_s}{2\Sigma_w^2}$.

- **S2**: the signal sample, $\mathbf{s}(n)$, has a known distribution. When $\mathbf{s}(n) \sim \mathcal{CN}(0, 2\sigma_s^2)$, thus

$$\mathbf{y}(n)|\mathbf{S2},\mathbf{f},\mathbf{g} \sim \mathcal{CN}\left(0, P_r\alpha|\mathbf{f}|^2|\mathbf{g}|^2 2\sigma_s^2 + 2\Sigma_w^2\right).$$

Here, $y_r(n)$ and $y_i(n)$ follow $\mathcal{N}\left(0, P_r\alpha|\mathbf{f}|^2|\mathbf{g}|^2\sigma_s^2 + \Sigma_w^2\right)$. Since Λ is a sum of $2N$ squares of i.i.d. Gaussian random variables with zero mean, Λ follows central chi-square distribution which is given as

$$f_{\Lambda|\mathcal{H}_1,\mathbf{S2}}(x) = \frac{x^{N-1} e^{-\frac{x}{(1+\gamma_e)2\Sigma_w^2}}}{\left((1+\gamma_e)2\Sigma_w^2\right)^N \Gamma(N)} \tag{5.20}$$

where $\gamma_e = \frac{P_r\alpha|\mathbf{f}|^2|\mathbf{g}|^2 2\sigma_s^2}{2\Sigma_w^2}$. Note that, in general, we can write $\gamma_e = \frac{P_r\alpha|\mathbf{f}|^2|\mathbf{g}|^2 P_s}{2\Sigma_w^2}$.

- **S3**: the signal sample, $\mathbf{s}(n)$, has an unknown distribution, but with known mean and variance. Since the distribution of $\mathbf{s}(n)$ is unknown, $y_r(n)$ and $y_i(n)$ also have unknown distributions, and thus the exact $f_{\Lambda|\mathcal{H}_1,\mathbf{S3}}(x)$ cannot be derived. However, $f_{\Lambda|\mathcal{H}_1,\mathbf{S3}}(x)$ can be derived approximately by using the CLT.

If $P_s = 2\sigma_s^2 = \frac{1}{N}\sum_{n=1}^{N}|s(n)|^2$, in general, the first-hop SNR can be given as $\gamma_1 = \frac{|\mathbf{f}|^2 P_s}{2\sigma_w^2}$ because $\gamma_1 = \frac{|\mathbf{f}|^2 \frac{1}{N}\sum_{n=1}^{N}|s(n)|^2}{2\sigma_w^2}$ for **S1** signal model, and $\gamma_1 = \frac{|\mathbf{f}|^2 2\sigma_s^2}{2\sigma_w^2}$ for

S2 and **S3** signal models. Similarly, the second-hop SNR can be given as $\gamma_2 = \frac{|g|^2 P_r}{2\sigma_w^2}$. Thus, the end-to-end SNRs for the non-coherent and coherent power coefficients can be given as

$$\gamma_e = \frac{\gamma_1 \gamma_2}{\gamma_2 + \epsilon} \quad \text{and} \quad \gamma_e = \frac{\gamma_1 \gamma_2}{\gamma_1 + \gamma_2 + 1}$$

respectively, where $\epsilon = 1 + \frac{P_s \mathbb{E}[|f|^2]}{2\sigma_w^2}$. The effective noise variances for the non-coherent and coherent power coefficients can also be given as

$$2\Sigma_w^2 = \left(\frac{\gamma_2 + \epsilon}{\epsilon}\right) 2\sigma_w^2 \quad \text{and} \quad 2\Sigma_w^2 = \left(\frac{\gamma_1 + \gamma_2 + 1}{\gamma_1 + 1}\right) 2\sigma_w^2$$

respectively.

Since distributions of the test statistic in (5.18)–(5.20) are similar to those of non-cooperative cases in (2.10), (2.12), (2.14), false-alarm and detection probabilities can be derive as (2.11), (2.13), (2.15) but with different effective noise variances and SNRs as defined above.

5.3.1.2 CLT Approach

If a sufficiently large number of samples are available, the distribution of Λ is close to a normal distribution such as $\Lambda \sim \mathcal{N}\left(\sum_{n=1}^{N} \mathbb{E}[\delta(n)], \sum_{n=1}^{N} \text{Var}[\delta(n)]\right)$. If it is possible to evaluate $\mathbb{E}[\delta(n)]$ and $\text{Var}[\delta(n)]$, the CLT approach can be applied to any signal model. The mean and variance are given below:

$$\mathbb{E}[\delta(n)] = \begin{cases} 2\Sigma_w^2 & : \mathcal{H}_0 \\ 2\Sigma_w^2(1 + \gamma_e) & : \mathbf{S1}, \mathbf{S2}, \mathbf{S3}. \end{cases}$$

$$\text{Var}[\delta(n)] = \begin{cases} \left(2\Sigma_w^2\right)^2 & : \mathcal{H}_0 \\ \left(2\Sigma_w^2\right)^2 (1 + 2\gamma_e) & : \mathbf{S1} \\ \left(2\Sigma_w^2\right)^2 (1 + \gamma_e)^2 & : \mathbf{S2} \\ P_r \alpha |f|^4 |g|^4 \left(\mathbb{E}[|s(n)|^4] - 4\sigma_s^4\right) + \left(2\Sigma_w^2\right)^2 (1 + 2\gamma_e) & : \mathbf{S3}. \end{cases}$$

For **S3**, if $s(n)$ is complex PSK, $\mathbb{E}[|s(n)|^4] = 4\sigma_s^4$, and thus $\text{Var}[\delta(n)] = \left(2\Sigma_w^2\right)^2 (1 + 2\gamma_e)$. By using CLT, the distribution of Λ can be given for i.i.d. samples as

$$\Lambda \sim \mathcal{N}\left(N\mathbb{E}[\delta(n)], N\text{Var}[\delta(n)]\right). \tag{5.21}$$

5.3.1.3   Multiple Cognitive Relays

A multiple-cognitive relay network is shown in Fig. 5.8. There are K cognitive relays between the primary user and the fusion center. The communication channels between cognitive relays and fusion center can be assumed as orthogonal to each other (e.g., based on time or frequency division multiple access (TDMA or FDMA) [5]). In orthogonal channels, the fusion center receives independent signals from cognitive relays. The fusion center is equipped with an energy detector. If MRC is considered at the fusion center, the CSI of channels of the first and second hops should be forwarded to the fusion center. However, CSI may not be available for energy detection which is non-coherent detection technique. In contrast to MRC, a receiver with SLC does not need instantaneous CSI of the channels in either hop, and consequently results in a low complexity system.[1] If the primary user is close to the fusion center, the fusion center can have a strong direct link from the primary user. The direct signal can also be combined at the SLC together with the relayed signals.

For i.i.d. relay channels, the output of the combiner forms the new decision statistic as

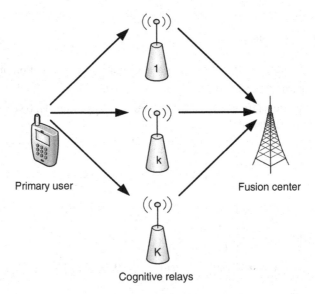

Fig. 5.8 Illustration of a multiple-cognitive relay network

---

[1]However, CSI of channels in the first hop is needed at the corresponding relays if coherent power coefficients are used.

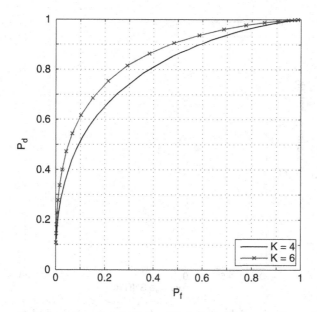

Fig. 5.9 ROC curves of **S1** signal model over Rayleigh fading channels for a multiple-cognitive relay network with the coherent power coefficient, $K = 4, 6, 2\sigma_w^2 = 1, N = 10$ and $\bar{\gamma} = 0\,\text{dB}$

$$\Lambda_{CSS} = \sum_{k=1}^{K} \Lambda_k = \sum_{k=1}^{K} \sum_{n=1}^{N} |\mathbf{y}_k(n)|^2$$

where $\mathbf{y}_k(n)$ and Λ_k are the received signal and test statistic of the kth cognitive relay, respectively. Under $\mathcal{H}_0$, Λ_{CSS} follows the same distribution as (5.18) but with KN degree of freedom. Similarly, under $\mathcal{H}_1$, Λ_{CSS} follows the same distribution as (5.19) or (5.20) for **S1** or **S2** signal model, respectively, but with KN degree of freedom and $\gamma_e = \sum_{k=1}^{K} \gamma_k$ where γ_k is the receive SNR from the kth cognitive relay.

Figure 5.9 shows ROC curves of a multiple-cognitive relay network with the coherent power coefficient case, in which the detection performance increases as the number of relays increases. Figure 5.10 shows the missed-detection probability versus average SNR for the same network scenario in which the detection diversity gain changes from 1 to 3 when K varies from 1 to 3. Thus, the detection diversity gain is K, which is same as in traditional relay networks [1–3, 13, 14, 20].

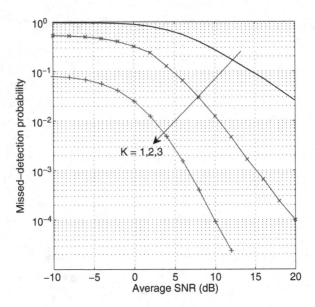

Fig. 5.10 The average missed-detection probability with average SNR of **S1** signal model over Rayleigh fading channels for a cognitive relay network with the coherent power coefficient, $K = 1, 2, 3, 2\sigma_w^2 = 1, N = 5$ and $\lambda = 10$

5.3.2 Decision Fusion

Each cognitive relay makes its own one-bit hard decision: '0' and '1' mean the absence and presence of primary activities, respectively, and the individual decisions are reported to the fusion center over a reporting channel (which can be with a narrow bandwidth). Capability of complex signal processing is needed at each cognitive relay. Two main assumptions are made for simplicity:

- The reporting channel is error-free [12, 19]
- The SNR statistics of the received primary signals are known at cognitive relays [22].

The fusion rule at the fusion center can be *OR, AND*, or *Majority* rule, which can be generalized as the "*m-out-of-K rule*". The decision device of the fusion center can be implemented with the m-out-of-K rule in which the fusion center decides on the presence of primary activity if there are m or more cognitive relays that individually decide on the presence of primary activity. When $m = 1, m = K$ and $m = \lceil K/2 \rceil$ (here $\lceil \cdot \rceil$ is the ceiling function), the m-out-of-K rule represents *OR rule, AND rule* and *Majority rule*, respectively. In the following, for simplicity of presentation, p_f and p_d are the false alarm and detection probabilities, respectively, for a cognitive relay, and P_f and P_d are the false alarm and detection probabilities, respectively, in the fusion center.

5.3.2.1 Reporting Channels Without Errors

If the sensing channels (the channels between the primary user and cognitive relays) are i.i.d., then every cognitive relay achieves identical false alarm probability p_f and detection probability p_d. If there are error free reporting channels (the channels between the cognitive relays and the fusion center), P_f and P_d at the fusion center can be written as

$$P_\chi = \sum_{i=m}^{K} \binom{K}{i} (p_\chi)^i (1 - p_\chi)^{K-i} \qquad (5.22)$$

where the notation 'χ' means 'f' or 'd' for false alarm or detection, respectively.

5.3.2.2 Reporting Channels with Errors

Because of the imperfect reporting channels, errors occur on the decision bits which are transmitted by the cognitive relays. Assume bit-by-bit transmission from cognitive relays. Thus, each identical reporting channel can be modeled as a binary symmetric channel (BSC) with cross-over probability p_e which is equal to the bit error rate (BER) of the channel.

Consider the ith cognitive relay. When the primary activity is present (i.e., under $\mathcal{H}_1$), the fusion center receives bit '1' from the ith cognitive relay when (1) the one-bit decision at the ith cognitive relay is '1' and the fusion center receives bit '1' from the reporting channel of the ith relay, with probability $p_d(1 - p_e)$; or (2) the one-bit decision at the ith cognitive relay is '0' and the fusion center receives bit '1' from the reporting channel of the ith relay, with probability $(1 - p_d)p_e$. On the other hand, when the primary activity is absent (i.e., under $\mathcal{H}_0$), the fusion center receives bit '1' from the ith cognitive relay when (1) the one-bit decision at the ith cognitive relay is '1' and the fusion center receives bit '1' from the reporting channel of the ith relay, with probability $p_f(1 - p_e)$; or (2) the one-bit decision at the ith cognitive relay is '0' and the fusion center receives bit '1' from the reporting channel of the ith relay, with probability $(1 - p_f)p_e$. Therefore, the overall false alarm and detection probabilities with the reporting error can be evaluated as

$$P_\chi = \sum_{i=m}^{K} \binom{K}{i} (p_{\chi,e})^i (1 - p_{\chi,e})^{K-i} \qquad (5.23)$$

where $p_{\chi,e} = p_\chi(1 - p_e) + (1 - p_\chi)p_e$ is the equivalent false alarm ('χ' is 'f') or detection ('χ' is 'd') probabilities of a cognitive relay.

Figures 5.11 and 5.12 show the ROC curves for m-out-of-K rule in decision fusion strategy for error-free and erroneous reporting channels, respectively. As Fig. 5.11, with error-free reporting channels, *OR* rule always outperforms *AND* and *Majority* rules, and *Majority* rule has better detection capability than

AND rule. As Fig. 5.12, with erroneous reporting channels, the comparative performances of the three fusion rules are not as clearcut. However, *OR* rule outperforms *AND* and *Majority* rules in lower detection threshold values. With the erroneous reporting channels, it cannot be expected $(P_f, P_d) = (1, 1)$ at $\lambda = 0$ and $(P_f, P_d) \to (0, 0)$ when $\lambda \to \infty$ on the ROC plot. When $\lambda = 0$, $P_f = P_d = \sum_{i=m}^{K} \binom{K}{i}(1 - p_e)^i p_e^{K-i}$; and when $\lambda \to \infty$, P_f and P_d approach $\sum_{i=m}^{K} \binom{K}{i} p_e^i (1 - p_e)^{K-i}$. In both scenarios, the values of P_d and P_f depend only on the error probabilities of the reporting channels. Figure 5.13 shows that *OR* rule has the maximum detection diversity order (i.e., 2 or 5 when $K = 2$ or 5), *AND* rule has worse detection diversity order which is always one, and *Majority* rule has detection diversity order $\lceil K/2 \rceil$ (e.g., 3 in Fig. 5.13 when $K = 5$).

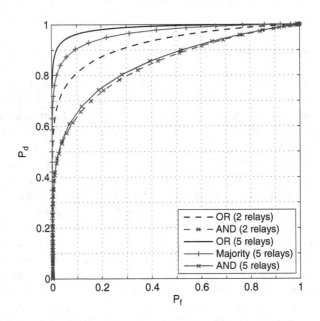

Fig. 5.11 ROC curves of **S1** signal model over Rayleigh fading channels for a multiple-cognitive relay network with decision fusion for *OR*, *AND* and *Majority* fusion rules with error-free reporting channels, $K = 2, 5, 2\sigma_w^2 = 1, N = 10$ and $\bar{\gamma} = 0$ dB

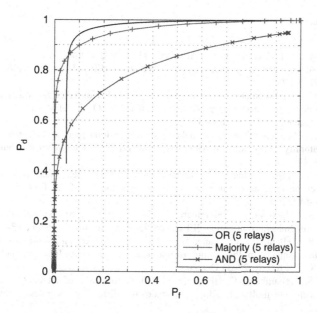

Fig. 5.12 ROC curves of **S1** signal model over Rayleigh fading channels for a multiple-cognitive relay network with decision fusion for *OR*, *AND* and *Majority* fusion rules with error in reporting channels, $p_e = 0.01$, $K = 5$, $2\sigma_w^2 = 1$, $N = 10$ and $\bar{\gamma} = 0\,\mathrm{dB}$

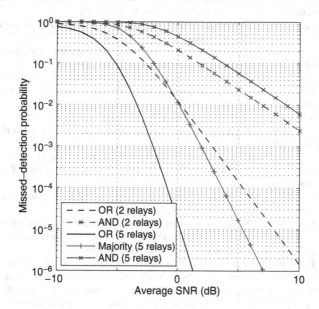

Fig. 5.13 The average missed-detection probability with average SNR of **S1** signal model over Rayleigh fading channels for a cognitive relay network with decision fusion for *OR*, *AND* and *Majority* fusion rules with error-free reporting channels, $K = 2, 5$, $2\sigma_w^2 = 1$, $N = 10$ and $\lambda = 10$

References

1. Amarasuriya, G., Tellambura, C., Ardakani, M. (2010) Output-threshold multiple-relay-selection scheme for cooperative wireless networks. IEEE T on Vehicular Technology **59**(6): 3091–3097.
2. Amarasuriya, G., Tellambura, C., Ardakani, M. (2011) Performance analysis framework for transmit antenna selection strategies of cooperative MIMO AF relay networks. IEEE T on Vehicular Technology **60**(7): 3030–3044.
3. Amarasuriya, G., Tellambura, C., Ardakani, M. (2012) Joint relay and antenna selection for dual-hop amplify-and-forward MIMO relay networks. IEEE T on Wireless Communications **11**(2): 493–499.
4. Atapattu, S., Jing, Y., Jiang, H., Tellambura, C. (2013) Relay selection schemes and performance analysis approximations for two-way networks. IEEE T on Communications **61**(3): 987–998.
5. Atapattu, S., Rajatheva, N., Tellambura, C. (2010) Performance analysis of TDMA relay protocols over Nakagami-m fading. IEEE T on Vehicular Technology **59**(1): 93–104.
6. Atapattu, S., Tellambura, C., Jiang, H. (2009) Energy detection of primary signals over $\eta - \mu$ fading channels. In: Proceedings of International Conference Industrial and Information Systems (ICIIS), Kandy, 28–31 Dec 2009.
7. Atapattu, S., Tellambura, C., Jiang, H. (2010) Performance of an energy detector over channels with both multipath fading and shadowing. IEEE T on Wireless Communications **9**(12): 3662–3670.
8. Digham, F. F., Alouini, M. S., Simon, M. K. (2003) On the energy detection of unknown signals over fading channels. In: Proceedings of IEEE International Conference on Communications (ICC), Anchorage, 11–15 May 2003.
9. Digham, F. F., Alouini, M. S., Simon, M. K. (2007) On the energy detection of unknown signals over fading channels. IEEE T on Communications **55**(1): 21–24.
10. Fan, R., Jiang, H. (2010) Optimal multi-channel cooperative sensing in cognitive radio networks. IEEE T on Wireless Communications **9**(3): 1128–1138.
11. Ganesan, G., Li, Y. (2007) Cooperative spectrum sensing in cognitive radio, part I: Two user networks. IEEE T on Wireless Communications **6**(6): 2204–2213.
12. Ghasemi, A., Sousa, E. S. (2005) Collaborative spectrum sensing for opportunistic access in fading environments. In: Proceedings of IEEE Dynamic Spectrum Access Networks (DySPAN), Maryland, 8–11 Nov 2005.
13. Gong, X., Vorobyov, S. A., Tellambura, C. (2011) Optimal bandwidth and power allocation for sum ergodic capacity under fading channels in cognitive radio networks. IEEE T on Signal Processing **59**(4): 1814–1826.
14. Gong, X., Vorobyov, S. A., Tellambura, C. (2011) Joint bandwidth and power allocation with admission control in wireless multi-user networks with and without relaying. IEEE T on Signal Processing **59**(4): 1801–1813.
15. Hasna, M. O., Alouini, M. S. (2004) A performance study of dual-hop transmissions with fixed gain relays. IEEE T on Wireless Communications **3**(6): 1963–1968.
16. Herath, S. P., Rajatheva, N. (2008) Analysis of equal gain combining in energy detection for cognitive radio over Nakagami channels. In: Proceedings of IEEE Global Telecommunications Conference (GLOBECOM), New Orleans, 30 Nov-4 Dec 2008.
17. Herath, S. P., Rajatheva, N., Tellambura, C. (2009) On the energy detection of unknown deterministic signal over Nakagami channels with selection combining. In: Canadian Conference on Electrical and Computing Engineering (CCECE), Newfoundland, 3–6 May 2009.
18. Laneman, J. N., Tse, D. N. C., Wornell, G. W. (2004) Cooperative diversity in wireless networks: Efficient protocols and outage behavior. IEEE T on Information Theory **50**(12): 3062–3080.

19. Mishra, S. M., Sahai, A., Brodersen, R. W. (2006) Cooperative sensing among cognitive radios. In: Proceedings of IEEE International Conference on Communications (ICC), Istanbul, 11–15 June 2006.

20. Ngo, D. T., Tellambura, C., Nguyen, H. H. (2009) Efficient resource allocation for OFDMA multicast systems with spectrum-sharing control. IEEE T on Vehicular Technology **58**(9): 4878–4889.

21. Papoulis, A., Pillai, S. U. (2002) Probability, Random Variables and Stochastic Processes, McGraw-Hill Companies, Inc.

22. Shen, J., Jiang, T., Liu, S., Zhang, Z. (2009) Maximum channel throughput via cooperative spectrum sensing in cognitive radio networks. IEEE T on Wireless Communications **8**(10): 5166–5175.

23. Zhang, W., Letaief, K. B. (2009) Cooperative communications for cognitive radio networks. Proceedings of the IEEE **97**(5): 878–893.